Student Solutions Manual

Third Edition

PREALGEBRA

VAN DYKE ▪ ROGERS ▪ ADAMS

Sharon Edgmon

Bakersfield College

Saunders College Publishing

Harcourt Brace College Publishers

Fort Worth Philadelphia San Diego New York Orlando Austin
San Antonio Toronto Montreal London Sydney Tokyo

Printed in the United States of America

ISBN 0-03-019952-2

890 129 765432

Preface

I would like to thank Jim Van Dyke, textbook author, and Carol Loyd, Developmental Editor, for their assistance in producing this manual, which consists of "worked-out" solutions to all odd-numbered exercises in the problem sets, and both even- and odd-numbered solutions for the True--False Concept Reviews and Chapter Tests.

This book is written in memory of my:
        daughter: Leah Edgmon
        father: Frans Johnson
        grandparents: William and Hilda Westlund

Sharon Edgmon
Professor of Mathematics
Bakersfield College
1801 Panorama Drive
Bakersfield, CA 93305

Table of Contents

CHAPTER ONE
WHOLE NUMBERS

Section 1.1
Whole Numbers: Writing, Rounding, and Inequalities

1. five hundred forty-two = 542

3. eight hundred ninety = 890

5. seven thousand, fifteen = 7015

7. 57 = fifty-seven

9. 7,500 = seven thousand, five hundred

11. 10,000,000 = ten million

13. twenty-five thousand, three hundred ten = 25,310

15. two hundred five thousand, three hundred ten = 205,310

17. forty-five million = 45,000,000

19. 243,700 = two hundred forty-three thousand, seven hundred

21. 23,470 = twenty-three thousand, four hundred seventy

23. 17,000,000 = seventeen million

25. 18 < 21

27. 51 > 44

29. 145 < 152

31. 348 < 351

33. 694 rounded to the ten's place = 690

35. 1658 rounded to the hundred's place is 1700

	Number	Ten	Hundred	Thousand	Ten Thousand
37.	102,385	102,390	102,400	102,000	100,000
39.	7,250,978	7,250,980	7,251,000	7,251,000	7,250,000

41. 560,353,730 = five hundred sixty million, three hundred fifty-three thousand, seven hundred thirty

43. one million, two hundred thirty-five thousand, nine hundred fifty-six = 1,235,956

45. 5634 < 5637

47. The smallest four-digit number is 1000.

49. 43,784,675 rounded to the ten thousand's place is 43,780,000.

51. 42,749 rounded to the nearest hundred is 42,700.
42,749 rounded to the nearest ten is 42,750.
42,750 rounded to the nearest hundred is 42,800.
We obtained a different result the second time because we rounded a previously rounded number.   The first method is the correct way to round the original number to the nearest hundred.

53. eighteen thousand, four hundred sixty-five dollars = $18,465

55. three hundred eighty-nine thousand, five hundred = 389,500

57. $36,407 = thirty-six thousand, four hundred seven dollars

59. $686,000 = six hundred eighty-six thousand dollars

61. 327,000,000 lb. = 327 million lb. by metal industries

63. twenty thousand, five hundred twenty-seven dollars = $20,527

65. Maine has the smallest per capita personal income.

67. 92,875,328 miles rounded to the nearest million miles is 93,000,000 miles.

69. 500,000 HIV infections in Western Europe.

71. The numbers in the table are rounded to the digit that is located in the position of largest place value for each piece of data.

73. 52,425,000 sq. mi. rounded to the nearest million sq. mi. is 52,000,000 sq. mi.

75. answers vary

77. The place value of the digit 4 is hundred billion.

79. 1145, 1229, 1234, 1243, 1324, 1342, 1432

81. 8275 rounded to the nearest ten thousand is 10,000.

83. answers vary

Section 1.2
Adding and Subtracting Whole Numbers

1. $65 + 132 = 97$

3. $748 + 231 = 979$

5. $\begin{array}{r} 756 \\ +\,236 \\ \hline 992 \end{array}$

7. We must carry the "1" from the 14 to the tens column.

9. $586 + 3492 + 321 = 4399$

11. $30,000 + 30,803 + 8740 = 69,543$

13. $8 + 90 + 403 + 6070 = 6571$

15. $2795 + 3643 + 7055 + 4004 = 17,497$ (rounded to the nearest hundred $= 17,500$)

17. $300 + 800 = 1100$

19. $7000 + 4000 = 11,000$

21. $3000 + 2000 + 5000 = 10{,}000$

23. estimate: $3000 + 7000 + 4000 + 2000 + 3000 = 19{,}000$
    exact: $3209 + 7095 + 4444 + 2004 + 3166 = 19{,}918$

25. estimate: $50{,}000 + 30{,}000 + 60{,}000 + 20{,}000 = 160{,}000$
    exact: $45{,}902 + 33{,}333 + 57{,}700 + 23{,}653 = 160{,}588$

27.   4 hundreds + 9 tens + 3 ones
    $-$ 2 hundreds + 9 tens + 2 ones
         2 hundreds + 0 tens + 1 one

29. $608 - 82 = 526$

31. $689 - 238 = 451$

33. The value of the "borrowed 1" is <u>10</u> ones.

35. $745 - 392 = 353$

37. $741 - 583 = 158$

39. $800 - 378 = 422$

41. $8743 - 4078 = 4665$ (rounded to the nearest hundred is 4700)

43.   5687
    $-$ 3499
         2188

45. $700 - 400 = 300$

47. $5000 - 3000 = 2000$

49. $50{,}000 - 30{,}000 = 20{,}000$

51. estimate: $900 - 400 = 500$
    exact: $875 - 406 = 469$

53. estimate: $7000 - 3000 = 4000$
    exact: $6580 - 3217 = 3363$

55. 1837 Fords + 1483 Toyotas + 241 Lexuses = 3561 cars sold

57. 2000 Hondas $-$ 1837 Fords = 163 more Hondas sold

59. 2000 Hondas + 1837 Fords + 1483 Toyotas = 5320 total cars sold

61. 1007 Jeeps − 241 Lexuses = 766 fewer Lexuses sold

63. estimate: 2000 + 2000 + 1000 + 1000 +1000 +1000 + 0 + 0 = 8000 cars sold
    exact: 2000 + 1837 + 1483 + 1309 + 1007 + 868 + 361 + 241 = 9106 cars sold

65. 24 + 9382 + 5093 + 27,853 = 42,352 (rounded to the nearest hundred is 42,400)

67. 50,000 − 29,315 = 20,685

69. estimate: 20,000 + 0 + 10,000 + 10,000 + 0 = 40,000
    exact: 23,706 + 34 + 7561 + 9346 + 236 = 40,883

71. estimate: 200,000 − 200,000 = 0
    exact: 203,855 − 195,622 = 8233

73. 50,000 + 50,000 + 3,000,000 = 3,100,000 HIV infections for Asia

75. 10,000,000 + 100,000 = 10,100,000 HIV infections for Africa

77. gross sales for the week are $767,847

79. $345,720 − $21,995 = $323,725 difference between the largest and smallest daily
    gross sales

81. 18,856 + 29,432 + 143,877 + 408,148 = 600,313 estimated arrests for violent
    crimes

83. 408,148 − 143,877 = 264,271 more aggravated assaults than robberies

85. $123,675 + $457,000 + $89,050 + $312,885 + $210,560 = $1,193,170
    (rounded to the nearest hundred dollars is $1,193,200)

87. 345,760 − 178,550 = 167,210 gallons per minute

89. $460 + $983 + $565 + $10,730 = $12,738
    $25,875 − $12,738 = $13,137 (rounded to the nearest hundred dollars is $13,100)

91. Bryce Canyon: 8500 ft − 6600 ft = 1900 ft
    Grand Canyon: 8300 ft − 2500 ft = 5800 ft
    Zion: 7500 ft − 4000 ft = 3500 ft
    Grand Canyon has the greatest change in elevation: 5800 ft
    This change in elevation is greater than Zion's change in elevation by 2300 ft
      (5800 ft − 3500 ft = 2300 ft)

93. answers vary

95. answers vary

97. 7,247,195 − 2,804,053 = 4,443,142 = four million, four hundred forty-three
    thousand, one hundred forty-two

99. A = 3, B = 6, C = 8

5A68	5368
241	241
10A9	1039
B64C	6648

101. A = 7, B = 2, C = 3, D = 1

4A6B	4762
− C251	− 3251
15D1	1511

103.   14,657
         3766
       123,900
          569
        54,861
       346,780
       544,533 (rounded to the nearest hundred is 544,500)

        14,700
         3,800
       123,900
          600
        54,900
       346,800
       544,700

Section 1.3
Multiplying and Dividing Whole Numbers

1. $(32)(3) = 96$

3. $132 \times 5 = 660$

5. $45 \times 11 = 495$

7. The place value of the product of "5" and "3" is <u>thousands</u>.

9. $(42)(38) = 1596$

11. $(89)(32) = 2848$

13. $646 \times 45 = 29,070$

15. $(92)(145) = 13,340$ (rounded to the nearest hundred is 13,300)

17. $40(40) = 1600$

19. $200(80) = 16,000$

21. $400 \times 20 = 8000$

23. estimate: $400 \times 80 = 32,000$
     exact: $412 \times 84 = 34,608$

25. estimate: $700 \times 50 = 35,000$
     exact: $684 \times 47 = 32,148$

27. $497 \div 7 = 71$

29. $355 \div 5 = 71$

31. $51 \div 23 = 2 \, r5$

33. The division has a remainder when the last difference in the division is smaller than the <u>divisor</u> and is not zero.

35. $12,208 \div 4 = 3052$

37. $\dfrac{768}{32} = 24$

39. $675 \div 43 = 15\,r30$

41. $(62)(58) = 3596$

43. $46{,}113 \div 57 = 809$ (rounded to the nearest ten is 810)

45. $600 \div 60 = 10$

47. $4000 \div 40 = 100$

49. $600\overline{)30{,}000}$ with quotient 50

$$600\overline{)\,30{,}000\,}^{\,50}$$

51. $\dfrac{35000}{700} = 50$

53. estimate: $60{,}000 \div 100 = 600$
    exact: $59{,}602 \div 103 = 578\,r68$

55. estimate: $800{,}000 \div 400 = 2000$
    exact: $780{,}854 \div 436 = 1790\,r414$

57. $606 \times 415 = 251{,}490$

59. $157{,}281 \div 309 = 509$

61. estimate: $600(2000) = 1{,}200{,}000$
    exact: $(633)(2361) = 1{,}494{,}513$

63. estimate: $8{,}000{,}000 \div 400 = 20{,}000$
    exact: $6{,}784{,}821 \div 423 = 16{,}039\,r324$

65. The estimate for the gross receipts from the sale of the Accords is:
    $600{,}000 = 30(\$20{,}000)$

67. The estimate of the gross receipts from the sale of Preludes is:
    $200{,}000 = 10(\$20{,}000)$
    The actual gross receipts from the sale of Preludes is $228{,}550 = 14(\$16{,}325)$

69. The estimate of the taxes paid per return during week 2 is:
$20,000,000 ÷ 4000 = $5000

71. The actual taxes paid per return during week 3 are:
$48,660,040 ÷ 11,765 = $4136 (rounded to the nearest hundred dollars is $4100)

73. estimate: (20)(50)(40) = 40,000
exact: (24)(45)(36) = 38,880

75. estimate: $\dfrac{200,000}{1000} = 200$

exact: $\dfrac{229,367}{1216} = 188\,r759$

77. 17(134 salmon) = 2278 salmon

79. 1664 trees ÷ 13-acres = 128 trees/acre

81. 375(3)(3)(3)(3) = 30,375 bacteria

83. $347,875 ÷ 5 = $69,575

85. 31(320,450 gal) = 9,933,950 gal (rounded to the nearest thousand is 9,934,000 **gal**)

87. 32,278 resistors ÷ 14 = 2305 radios with 8 resistors left over

89. 450($78) = $35,100 is the cost of the radios
450($112) = $50,400 is the net income of the sale of the radios
$50,400 − $35,100 = $15,300 is the profit from the sale of the radios

91. $\dfrac{780 \text{ tons}}{12 \text{ tons}} \cdot 8 \text{ hr} = 520 \text{ hr}$

93. Latin America = 2 million;   North America = 1 million
North Africa = 100,000;   Eastern Europe = 50,000
Eastern Asia = 50,000;   Australia = 25,000
North America = 1 million;   Western Europe = 500,000
North Africa = 100,000;   Eastern Asia = 50,000
Eastern Europe = 500,000;   Australia = 25,000

95. $290,000,000 ÷ 36,000 = $8055 (rounded to the nearest ten dollars is $8060)

97. Los Angeles is getting a better deal than Portland by one-half ($\dfrac{\$16,000}{\$8000}$)

9

99. answers vary

101. answers vary

103. answers vary

105. 11,575(75 bushels) = 868,125 bushels
    868,125($27) = $23,439,375 (rounded to the nearest thousand $ is $23,439,000)

107. A = 6, B = 3, C = 8, E = 1

```
     51A              516
   ×  B2            ×  32
    10B2             1032
    154C             1548
   1A5E2            16512
```

109. A = 5, B = 1, C = 6

```
     5AB2             5512
  3)1653C          3)16536
```

111.
23 × 10 = 230	23 × 100 = 2300	23 × 1000 = 23,000
56 × 10 = 560	56 × 100 = 5600	56 × 1000 = 56,000
789 × 10 = 7890	789 × 100 = 78,900	789 × 1000 = 789,000
214 × 10 = 2140	214 × 100 = 21,400	214 × 1000 = 214,000
1345 × 10 = 13,450	1345 × 100 = 134,500	1345 × 1000 = 1,345,000

When multiplying a number by 10, move the decimal point one place to the right; by 100, two places to the right; by 1000, three places to the right.

23,000,000 ÷ 10 = 2,300,000	140,000,000 ÷ 10 = 14,000,000
23,000,000 ÷ 100 = 230,000	140,000,000 ÷ 100 = 140,000
23,000,000 ÷ 1000 = 2,300	140,000,000 ÷ 1000 = 14,000

When dividing a number by 10, move the decimal point one place to the left; by 100, two places to the left; by 1000, three places to the left.

Section 1.4
Whole Number Exponents and Powers of Ten

1. $12(12)(12)(12)(12)(12) = 12^6$ 3. $9^2 = 81$

5. $2^3 = 8$ 7. $19^0 = 1$

9. In $7^3 = 343$, 7 is the <u>base,</u> 3 is the <u>exponent,</u> and 343 is the <u>product.</u>

11. $6^3 = 216$ 13. $19^2 = 361$

15. $10^4 = 10,000$ 17. $8^3 = 512$

19. $3^8 = 6561$ 21. $45 \times 10^2 = 4500$

23. $7 \times 10^4 = 70,000$ 25. $1200 \div 10^2 = 12$

27. $340,000 \div 10^3 = 340$

29. ...write as many zeros to the right of the number as the <u>power</u> of ten.

31. $435 \times 10^4 = 4,350,000$ 33. $1,200,000 \div 10^3 = 1200$

35. $3591 \times 10^4 = 35,910,000$ 37. $\dfrac{30,200}{100} = 302$

39. $705 \times 10^8 = 70,500,000,000$ 41. $970,000,000 \div 10^5 = 9700$

43. $10(10)(10)(10)(10)(10)(10)(10)(10)(10)(10) = 10^{11}$

45. $14^4 = 38,416$ 47. $9^9 = 387,420,489$

49. $3350 \times 10^9 = 3,350,000,000,000$ 51. $\dfrac{438,000,000,00}{10^8} = 4380$

53. $\$73 \times 10^6 = \$73,000,000$ 55. $32 \times 10^6 = 32,000,000$

57. Sub-Saharan Africa: $10,000,000 = 10^7$
    North Africa/Middle East: $100,000 = 10^5$

59. West. Europe: $500,000 = 50,000 \times 10 = 5,000 \times 10^2 = 500 \times 10^3 = 50 \times 10^4 = 5 \times 10^5$

61. $\$5^9 = \$1,953,125$

63. $4^{10} = $ ten factors of 4

65. The deposit on Mitchell's twelfth birthday is $\$3^{12} = \$531,441$.
    The total amount deposited in the account is: $\$3 + \$9 + \$27 + \$81 + \$243 + \$729 + \$2187 + \$6561 + \$19,683 + \$59,049 + \$177,147 + \$531,441 = \$797,160$

67. $\left(4^3 + 8^3 + 11^3 + 23^3\right) + \left(2^4 + 5^4 + 6^3 + 7^4\right) =$
    $(64 + 512 + 1331 + 12,167) - (16 + 625 + 1296 + 2401) = 14,074 - 4338 = 9736$

Section 1.5
Operations with Exponents

1. $6^2 \cdot 6^4 = 6^{2+4} = 6^6$

3. $11^4 \cdot 11^6 = 11^{4+6} = 11^{10}$

5. $22^6 \cdot 22^2 = 22^{6+2} = 22^8$

7. $14^3 \cdot 14^8 = 14^{3+8} = 14^{11}$

9. $19^{11} \cdot 19^4 = 19^{11+4} = 19^{15}$

11. $58^7 \cdot 58^{19} = 58^{7+19} = 58^{26}$

13. $6^2 \cdot 6^4 \cdot 6^7 = 6^{2+4+7} = 6^{13}$

15. $\left(8^2\right)^2 = 8^{2 \cdot 2} = 8^4$

17. $\left(10^4\right)^3 = 10^{4 \cdot 3} = 10^{12}$

19. $\left(19^0\right)^{11} = 19^{0 \cdot 11} = 19^0$

21. $\left(14^5\right)^4 = 14^{5 \cdot 4} = 14^{20}$

23. $\left(12^6\right)^3 = 12^{6 \cdot 3} = 12^{18}$

25. $\left(10^3\right)^7 = 10^{3 \cdot 7} = 10^{21}$

27. $\left(18^5\right)^9 = 18^{5 \cdot 9} = 18^{45}$

29. $\left(2 \cdot 5\right)^2 = 2^2 \cdot 5^2$

31. $\left(8 \cdot 9\right)^4 = 8^4 \cdot 9^4$

33. $\left(13 \cdot 12\right)^7 = 13^7 \cdot 12^7$

35. $\left(12 \cdot 23\right)^{12} = 12^{12} \cdot 23^{12}$

37. $(14 \cdot 45)^{19} = 14^{19} \cdot 45^{19}$

39. $(4 \cdot 6 \cdot 8)^{11} = 4^{11} \cdot 6^{11} \cdot 8^{11}$

41. $(13^4)^7 = 13^{4 \cdot 7} = 13^{28}$

43. $16^3 \cdot 16^7 \cdot 16^8 = 16^{3+7+8} = 16^{18}$

45. $(24^7)^9 = 24^{7 \cdot 9} = 24^{63}$

47. $(3^3 \cdot 7^2)^4 = 3^{3 \cdot 4} \cdot 7^{2 \cdot 4} = 3^{12} \cdot 7^8$

49. $\left[(5^2)^3\right]^4 = 5^{2 \cdot 3 \cdot 4} = 5^{24}$

51. $V = s^3 = (5.)^3 = 125 \, \text{in}^3$

53. $V = lwh = w(w)(w^2) = w^4 = 5^4 = 625 \, \text{ft}^3$

55. $V = \pi r^2 h = \pi r^2 \cdot r^3 = \pi r^5 = 3 \cdot 3^5 = 3^6 = 729 \, \text{in}^3$

57. Sub Saharan Africa has 100 times the estimated number of HIV infections as in North Africa/Middle East.

   Sub Saharan Africa $= 10^7 = 10,000,000$ HIV infections

   North Africa/Middle East $= 10^5 = 100,000$ HIV infections

   $$10^5 \cdot 10^2 = 10^7 \qquad \text{or} \qquad \frac{10^7}{10^5} = 10^2$$

59. answers vary

61. If $(2^n)^4 = 256 = 2^8$, then $4n = 8$ or $n = 2$.

63. If $(2 \cdot 3)^n = 216$ and $6^n = 6^3$, then $n = 3$.

Section 1.6
Order of Operations and Averages

1. Simplify:

$5 \cdot 8 + 13$

$40 + 13$

53

3. Simplify:

$15 - 5 \cdot 3$

$15 - 15$

0

5. Simplify:

$24 + 6 \div 2$

$24 + 3$

27

7. Simplify:

$30 - (13 + 2)$

$30 - 15$

15

9. Simplify:

$30 \div 6 \times 5$

5×5

25

11. Simplify:

$21 + 5 \cdot 4 - 2$

$21 + 20 - 2$

$41 - 2$

39

13. Simplify:

$3^4 + 4^3$

$81 + 64$

145

15. Simplify:

$4 \cdot 7 + 3 \cdot 5$

$28 + 15$

43

17. Simplify:

$3^2 - 4 \cdot 2 + 5 \cdot 6$

$9 - 8 + 30$

31

19. Simplify:

$36 \div 9 + 8 - 5$

$4 + 8 - 5$

7

21. Simplify:

$(14 + 28) - (34 - 27)$

$42 - 7$

35

23. Simplify:

$49 \div 7 \cdot 3^3 + 7 \cdot 4$

$7 \cdot 27 + 28$

$189 + 28$

217

25. Simplify:

$$96 \div 12 \cdot 3$$
$$8 \cdot 3$$
$$24$$

27. Simplify:
$$72 - 4(19 - 10) + 11 - 19$$
$$72 - 4(9) + 11 - 19$$
$$72 - 36 + 11 - 19$$
$$36 + 11 - 19$$
$$47 - 19$$
$$28$$

29. $\dfrac{3+7}{2} = \dfrac{10}{2} = 5$

31. $\dfrac{8+10}{2} = \dfrac{18}{2} = 9$

33. $\dfrac{8+14+17}{3} = \dfrac{39}{3} = 13$

35. $\dfrac{3+5+7+9}{4} = \dfrac{24}{4} = 6$

37. $4(8) - (4 + 7 + 9) = 32 - 20 = 12$

39. $\dfrac{14+18+30+42}{4} = \dfrac{104}{4} = 26$

41. $\dfrac{25+35+45+55}{5} = \dfrac{160}{4} = 40$

43. $\dfrac{7+14+16+23+30}{5} = \dfrac{90}{5} = 18$

45. $\dfrac{8+11+19+28+44+76}{6} = \dfrac{186}{6} = 31$

47. $\dfrac{31+130+238+277}{4} = \dfrac{676}{4} = 169$

49. $\dfrac{101+105+108+126}{4} = \dfrac{440}{4} = 110$

51. $5(59) - (34 + 81 + 52 + 74) = 295 - 241 = 54$

53. Simplify:
$$50 - 12 \div 6 - 36 \div 6 + 3$$
$$50 - 2 - 6 + 3$$
$$48 - 6 + 3$$
$$42 + 3$$
$$45$$

55. $\dfrac{183+526+682+589+720}{5} = \dfrac{2700}{5} = 540$

57. mallards + canvasbacks = 990 + 740 = 1730
teal + woodducks = 924 + 210 = 1134
1730 − 1134 = 596 more mallards + canvasbacks than teal + woodducks

59. wooducks = 210;   four times the number of wooducks = 4(210) = 840
wooducks + mallards = 840 + 990 = 1830
teal + canavasbacks 924 + 740 = 1664
1830 − 1664 = 166 more wooducks + mallards than teal + canvasbacks

61. average number of the species of ducks = $\dfrac{990 + 924 + 740 + 210}{4} = \dfrac{2864}{4} = 716$

63. Simplify:

$11\left(3^3 \cdot 4 - 98\right) \div 5 - 21$

$11(27 \cdot 4 - 98) \div 5 - 21$

$11(108 - 98) \div 5 - 21$

$11(10) \div 5 - 21$

$110 \div 5 - 21$

$22 - 21$

1

65. Simplify:

$3(7 - 3)^3 - 8(3 - 1)^2$

$3(4)^3 - 8(2)^2$

$3(64) - 8(4)$

$192 - 32$

160

67. $\dfrac{3232 + 4343 + 5454 + 6565 + 7676 + 8790}{6} = \dfrac{36{,}060}{6} = 6010$

69. $\dfrac{112{,}315 + 236{,}700 + 156{,}865 + 103{,}674 + 300{,}071}{5} = \dfrac{909{,}625}{5} = 181{,}925$

71. $35(\$8) + 62(\$10) = \$280 + \$620 = \$900$

73. $\dfrac{255 + 198 + 210 + 300 + 193 + 213 + 278 + 200 + 205}{9} = \dfrac{2052}{9} = 228$

75. $1(\$25) + 2(\$18) = \$25 + \$36 = \$61$

77. $\dfrac{61 + 50 + 48 + 45 + 41}{5} = \dfrac{245}{5} = 49$ mph

79. net income for one camera = \$575 − \$346 = \$229
net income for one lens = \$316 − \$212 = \$104
net income for cameras and lenses: $27(\$229) + 19(\$104) = \$6183 + \$1976 = \$8159$

81. Sub-Saharan Africa = 10,000,000 HIV infections
North Africa/Middle East = 100,000 HIV infections
$\dfrac{10{,}000{,}000 + 100{,}000}{2} = \dfrac{10{,}100{,}000}{2} = 5{,}050{,}000$ HIV infections

16

83. The average price per can is:
$$\frac{3(84)+1(82)+4(78)+3(74)+2(71)+2(65)}{15}=$$
$$\frac{252+82+312+222+142+130}{15}=\frac{1140}{15}=76 \text{ cents/can}$$

85. The average score for the tournament is:
$$\frac{66+3(68)+7(69)+8(70)+12(71)+16(72)+15(74)+10(76)+6(78)+3(82)+85}{82}=$$
$$\frac{66+204+483+560+852+1152+1110+760+468+246+85}{82}=$$
$$\frac{5986}{82}=73$$

87. $3 \cdot 5 + 4 = 15 + 4 = 19$ is correct

89. Simplify:
$$2(36 \div 3^2 +1) - 2 + 12 \div 4(3)$$
$$2(36 \div 9 +1) - 2 + 12 \div 4(3)$$
$$2(4+1) - 2 + 3(3)$$
$$2(5) - 2 + 3(3)$$
$$10 - 2 + 9$$
$$8 + 9$$
$$17$$

91. Simplify:
$$(6 \cdot 3 - 8)^2 - 50 + 2 \cdot 3^2 + 2(9-5)^3$$
$$(18-8)^2 - 50 + 2 \cdot 9 + 2(4)^3$$
$$(10)^2 - 50 + 18 + 2(64)$$
$$100 - 50 + 18 + 128$$
$$50 + 18 + 128$$
$$68 + 128$$
$$196$$

93.
$$\frac{5(\$153)+13(\$125)+24(\$110)+30(\$100)+30(\$75)+24(\$50)+14(\$25)+10(\$17)}{150}=$$
$$\frac{\$765+\$1625+\$2640+\$3000+\$2250+\$1200+\$350+\$170}{150}=$$
$$\frac{\$12,000}{150}=\$80=\text{average}$$
The patron missed the average by $8 = $80 − $72
Therefore, her donation will be 150($8) = $1200

95. group activity

Section 1.7
Reading and Interpreting Tables

1. least expensive year in the '90's was 1992

3. 1990's total summer costs = $401 million + $456 million = $857 million

5. difference in costs between 2004 and 2002:
    $793 million − $545 million = $248 million

7. estimate: NBC rights from 1984 to 2008 (in millions) =
    $300 + $400 + $500 + $700 + $500 + $800 + $600 + $900 =$4700 million
    actual: $300 + $401 + $456 + $705 + $545 + $793 + $613 + $894 =$4707 million

9. Fish cakes has the highest level of cholesterol per serving.

11. Fish cakes cholesterol − chicken Dijon cholesterol = 147 mg− 99 mg = 48 mg

13 veal chop fat = 3(24 g) = 72 g

15. veal chop calories + chicken Dijon calories + pepper steak calories:
    421 cal + 247 cal + 240 cal = 908 cal

17. The three entrees are fish cakes, chicken Dijon and pepper steak:
    259 cal. + 247 cal. + 240 cal. = 746 cal. < 900 cal maximum

19. gain in account value from age 57 to age 65 = $251,206 −$144,276 = $106,930

21. the loss at age 59 = $165,219 − $150,928 = $14,291

23. death benefit − account value at age 61 = $400,000 − $189,522 = $210,478

25. the largest increase in account value occurred between the ages of 63 and 65

27. the most picnics are held in August

29. total number of overnight campers = 231 + 378 + 1104 + 1219 + 861 = 3793

31. July picnic income = 2371($5) = $11,855

33. August hikers − May hikers = 192 − 48 = 144

35. Fisher Zoo had the greatest increase in attendance from 1993 to 1996:
    $2,720,567 - 2,356,890 = 363,677$

37. estimate of the attendance for 5 years at the Fisher Zoo:
    $2,000,000 + 2,000,000 + 3,000,000 + 3,000,000 + 3,000,000 = 13,000,000$

39. average attendance for 5 years at the Delaney Zoo:
    $$\frac{1,067,893 + 1,119,875 + 1,317,992 + 1,350,675 + 1,398,745}{5} = \frac{6,255,180}{5} = 1,251,036$$

41. revenue for 1995 at the Fisher Zoo = $25(2,745,111) = $68,627,775$
    ($68,628,000 rounded to the nearest thousand dollars)

43. answers vary

45. The attendance in 1997 for Utaki Park to average 501,200 in attendance for the
    years 1992 to 1997: $6(501,200) - 2,408,855 = 3,007,200 - 2,408,855 = 598,345$

Section 1.8
Drawing and Interpreting Tables

1. The number of phone calls is the greatest from 10-11.

3. There are 250 phone calls made from 2-3.

5. The total number of phone calls is:
   $50 + 150 + 300 + 225 + 50 + 125 + 250 + 200 + 75 = 1425$

7. There are 40 vans in the shop for repair.

9. Full size cars are the most type of car in the repair shop.

11. There are 400 total cars in the shop for repair ($80 + 160 + 40 + 120$)

13. 1996 is the greatest production year.

15. The increase in production between 1993 and 1994 is: $22,500 - 10,000 = 12,500$

17. The average production per year is: $\dfrac{15+10+22.5+25+30}{5} = \dfrac{102.5}{5} = 20,500$

19. paint + lumber = $5000 + $40,000 = $4500

21. plastics − steel castings = $35,000 − $20,000 = $15,000

23. steel castings = 2($20,000) = $40,000

25. Distribution of grades in an algebra class:

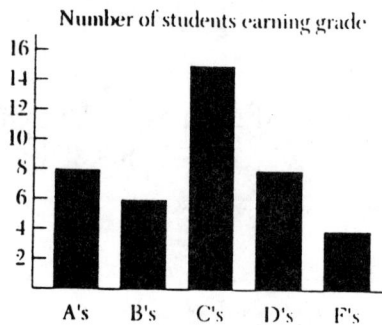

Number of students earning grade

27. Career performance as expressed by a senior class:

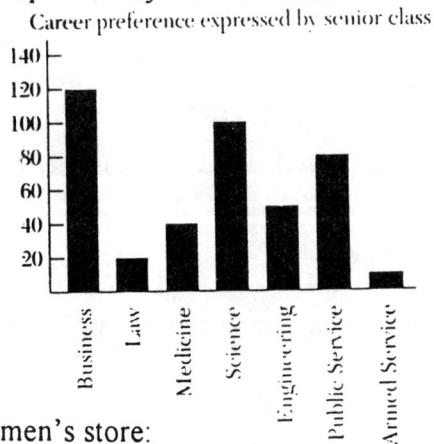

Career preference expressed by senior class

29. Daily sales at the local men's store:

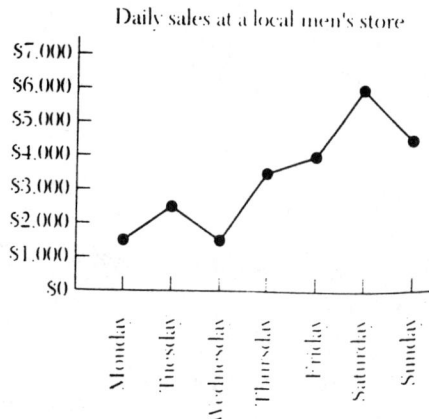

Daily sales at a local men's store

31. Income from various sources for a given year for the Smith family:

Annual sources of income for the Smith family

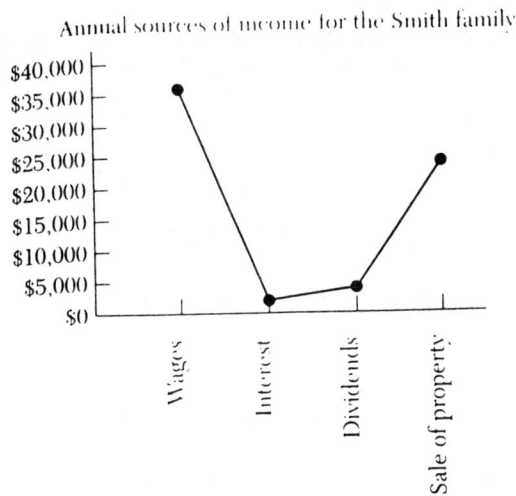

33. The cost of an average three-bedroom house in a rural city:

Year	Cost of an average three-bedroom house 🏠 = $10,000 🏠 = $5000
1970	🏠 🏠 🏠 🏠 🏠 🏠
1975	🏠 🏠 🏠 🏠 🏠 🏠 🏠
1980	🏠 🏠 🏠 🏠 🏠 🏠 🏠 🏠
1985	🏠 🏠 🏠 🏠 🏠 🏠 🏠
1990	🏠 🏠 🏠 🏠 🏠 🏠 🏠
1995	🏠 🏠 🏠 🏠 🏠 🏠 🏠 🏠

35. The oil production from a local well over a 5-year period:

Year	Barrels produced 🛢 = 5000 ⊏ = 2500
1991	🛢 🛢 🛢
1992	🛢 🛢 🛢 🛢 ⊏
1993	🛢 🛢 🛢 🛢 🛢 🛢 🛢
1994	🛢 🛢 🛢 🛢 🛢 🛢 ⊏
1995	🛢 🛢 🛢 🛢 🛢 🛢 🛢 🛢

37. answers vary

39. answers vary

Chapter 1
True-False Concept Review

1. True

2. True

3. False – The word name for 750 is seven hundred fifty. The word "and" represents the position of the decimal point.

4. False – 500 > 23

5. True

6. False – To the nearest ten, 7449 rounds to 7450.

7. False – The rounded value of a number may also be equal to the number.

8. True

9. True

10. False – The sum of 8 and 5 is 13.

11. True

12. True

13. True

14. True

15. True

16. False – The factors of 15 are 1, 3, 5 and 15.

17. True

18. True

19. True

20. True

21. False - $\dfrac{2,200,000}{100,000} = 220$

22. False – In the order of operations multiplication and division are done in the order that they appear from left to right.

23. False – In the order of operations, multiplication is always done before addition unless grouping symbols are present to indicate otherwise.

24. True

25. True

Chapter 1
Review

1. 892 = eight hundred ninety-two

3. 680,057 = six hundred eighty thousand, fifty-seven

5. two hundred eight million, twenty-five thousand, six hundred eight = 208,025,608

7. 48 < 62

9. 65,007 > 60,005

11. 4769 rounded to the nearest ten is 4770

13. 67,349 rounded to the nearest hundred is 67,300

15. 3,044,999 rounded to the nearest ten thousand is 3,040,000

17. 843 + 629 + 1208 + 77 + 45 = 2802

19. 1,305,202 + 126,433 + 805 + 65,577 + 43 = 1,498,060

21. 600 + 900 + 500 + 500 = 2500

23. 1000 + 900 + 700 + 700 + 400 = 3700

25. $50{,}000 + 60{,}000 + 60{,}000 + 70{,}000 + 80{,}000 = 320{,}000$

27. $91{,}211 - 3368 = 87{,}843$

29. $50{,}008 - 30{,}684 = 19{,}324$

31. $900 - 700 = 200$

33. $30{,}000 - 20{,}000 = 10{,}000$

35. $80{,}000 - 40{,}000 = 40{,}000$

37. $54(189) = 10{,}206$

39. $(6)(21)(394) = 49{,}644$

41. $40(100) = 4000$

43. $700(400) = 280{,}000$

45. $20(700)(200) = 2{,}800{,}000$

47. $3{,}293{,}988 \div 482 = 6834$

49. $15{,}440 \div 225 = 68 \ r140$

51. $30{,}000 \div 500 = 60$

53. $60{,}000 \div 3000 = 20$

55. $900{,}000 \div 900 = 1000$

57. $12^3 = 12 \cdot 12 \cdot 12 = 1728$

59. $5^4 = 5 \cdot 5 \cdot 5 \cdot 5 = 625$

61. $340 \times 10^5 = 34{,}000{,}000$

63. $2{,}700{,}000 \div 10^3 = 2700$

65. $39 \times 10^{12} = 39{,}000{,}000{,}000{,}000$

67. $11^4 \cdot 11^5 = 11^{4+5} = 11^9$

69. $31^2 \cdot 31^2 \cdot 31^4 = 31^{2+2+4} = 31^8$

71. $\left(7^5\right)^2 = 7^{5 \cdot 2} = 7^{10}$

73. $\left(12^2\right)^6 = 12^{2 \cdot 6} = 12^{12}$

75. $\left(22^6\right)^7 = 22^{6 \cdot 7} = 22^{42}$

77. $(9 \cdot 5)^6 = 9^6 \cdot 5^6$

79. $(4 \cdot 3)^{10} = 4^{10} \cdot 3^{10}$

81. Simplify:

$$25 - 60 \div 12 \times 3$$
$$25 - 5 \times 3$$
$$25 - 15$$
$$10$$

83. Simplify:
$$95 - 2 \cdot 4^2 \div 8 + 9$$
$$95 - 2 \cdot 16 \div 8 + 9$$
$$95 - 32 \div 8 + 9$$
$$95 - 4 + 9$$
$$91 + 9$$
$$100$$

85. Simplify:

$144 \div 8 + 2^3$

$18 + 8$

26

87. $\dfrac{78 + 92 + 103 + 129 + 143}{5} = \dfrac{545}{5} = 109$

89. $\dfrac{8 + 37 + 125 + 48 + 117 + 193}{6} = \dfrac{528}{6} = 88$

91. $480 + 250 + 270 = 1000$ units

93. The Day shift had the highest production per day.

Day: $480 \div 20 = 24$ Swing: $250 \div 20 = 12.5$ Graveyard: $270 \div 18 = 15$
Day: $480 \div 80 = 6$ Swing: $250 \div 50 = 5$

95. The value of the graveyard shift's production is $270(\$1495) = \$403{,}650$.
(rounded to the nearest thousand dollars is $404,000)

97. Vans had the least sales.

99. The expected number of 2-door sedans sold is: $2(25) = 50$ **vehicles**

101. The daily gross sales at the local dairy:

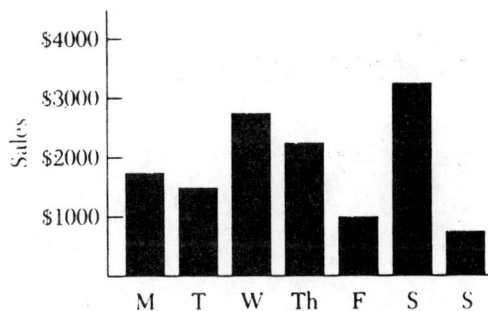

103. Disposition of family income:

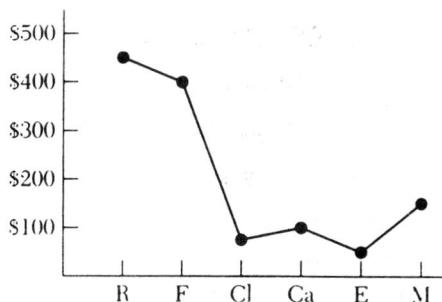

Chapter 1
Test

1. $54\overline{)5886}$ with quotient 109

2. $9123 - 6844 = 2279$

3. Simplify:
$$36 \div 9 + 4 \cdot 5 - 5$$
$$4 + 20 - 5$$
$$24 - 5$$
$$19$$

4. $53(768) = 40{,}704$

5. $278 > 201$

6. $76 \times 10^4 = 760{,}000$

7. $709(386) = 273{,}674$

8. four hundred fifty thousand, eighty-two $= 450{,}082$

9. $\dfrac{1294 + 361 + 1924 + 274 + 682}{5} = \dfrac{4535}{5} = 907$

10. $35(2095) = 73{,}325$ (rounded to the nearest hundred is $73{,}300$)

11. $17{,}852$ rounded to the nearest hundred is $17{,}900$

12. $80{,}000 + 20{,}000 + 50{,}000 + 10{,}000 = 160{,}000$

13. $9^3 = 9 \cdot 9 \cdot 9 = 729$

14. $39 + 953 + 4 + 4886 = 5882$

15. $43^8 \cdot 43^5 = 43^{8+5} = 43^{15}$

16. $7040 - 587 = 6453$

17. $6007 =$ six thousand, seven

18. Simplify:
$$25 + 2^3 - 24 \div 8$$
$$25 + 8 - 3$$
$$33 - 3$$
$$30$$

19. Add:
$$
\begin{array}{r}
45{,}974 \\
31{,}900 \\
78{,}211 \\
12{,}099 \\
67{,}863 \\
\hline
236{,}047
\end{array}
$$

20. $6{,}050{,}000{,}000 \div 10^5 = 60{,}500$

21. 524,942,664 rounded to the nearest ten thousand is 524,940,000

22. estimate: $48,000 \div 80 = 600$ actual: $47,125 \div 76 = 620 \, r5$

23. Simplify:

    $72 - 8^2 + 27 \div 3$

      $72 - 64 + 9$

        $8 + 9$

         17

24. Simplify:

    $(4 \cdot 2)^3 + (3^2)^2 + 9 \cdot 2$

      $8^3 + 9^2 + 18$

      $512 + 81 + 18$

        611

25. $\dfrac{582 + 678 + 425 + 979}{4} = \dfrac{2664}{4} = 666$

26. (700 words)(12 pages) = 8400 words $\div$ 80 words/minute = 105 minutes

27. $\$9,456,000 \div 12 = \$788,000;$ $\$788,000 \div 20 = \$39,400$

28. a. Chevrolet had the greatest sales.
    b. Pontiacs + Chevrolets = 200 + 500 = 700 automobiles
    c. Buicks – Oldsmobiles = 400 – 300 = 100 automobiles

29. a. Division A and Division B each have the same number of employees.
      Division A = 525   Division B = 525   Division C = 150
    b. Division A Day shift – Division B Day shift = 350 – 125 = 225 employees
    c. Total employees in the three divisions = 525 + 525 + 150 = 1200 employees

30. Lunches purchased at the local fast food bar over one week:

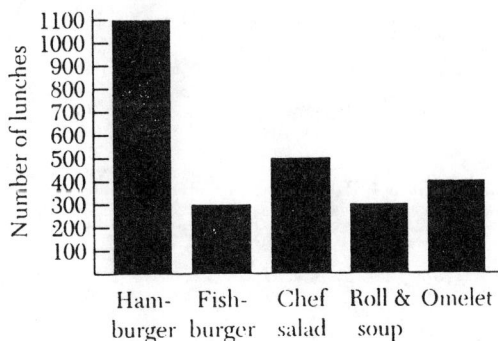

CHAPTER TWO
MEASUREMENT

Section 2.1
English and Metric Measurement

1. $(4 \text{ ft}) \cdot 6 = 24 \text{ ft}$

3. $(200 \text{ ml}) \div 25 = 8 \text{ ml}$

5. $(80 \text{ gal}) \div 20 = 4 \text{ gal}$

7. $3 \cdot (317 \text{ oz}) = 951 \text{ oz}$

9. $(400 \text{ hrs}) \div 8 = 50 \text{ hr}$

11. $(2912 \text{ lbs}) \div 14 = 208 \text{ lbs}$

13. $(56 \text{ sec}) \cdot (20) = 1120 \text{ sec}$

15. $6 \text{ lb} + 14 \text{ lb} = 20 \text{ lb}$

17. $5 \text{ yd} + 8 \text{ yd} + 4 \text{ yd} = 17 \text{ yd}$

19. $32 \text{ g} - 12 \text{ g} = 20 \text{ g}$

21. $360 \text{ kl} - 155 \text{ kl} = 205 \text{ kl}$

23. $48 \text{ mm} + 32 \text{ mm} + 10 \text{ mm} = 90 \text{ mm}$

25. $321 \text{ yd} - 217 \text{ yd} = 104 \text{ yd}$

27. $624 \text{ gal} - 209 \text{ gal} + 138 \text{ gal} = 553 \text{ gal}$

29. $210 \text{ cm} - 45 \text{ cm} + 24 \text{ cm} - 165 \text{ cm} = 24 \text{ cm}$

31. answers vary

33. $(6 \text{ ft } 5 \text{ in}) + (2 \text{ ft } 7 \text{ in}) + (10 \text{ ft } 7 \text{ in}) = 18 \text{ ft } 19 \text{ in} = 19 \text{ ft } 7 \text{ in}$

35. $(21 \text{ min } 39 \text{ sec}) - (14 \text{ min } 47 \text{ sec}) = 6 \text{ min } 52 \text{ sec}$

37. $(35 \text{ min } 12 \text{ sec}) \cdot 6 = 210 \text{ min } 72 \text{ sec} = 3 \text{ hr } 31 \text{ min } 12 \text{ sec}$

39.
$$
\begin{array}{r}
2 \text{ yd } 2 \text{ ft } 6 \text{ in} \\
+ \ \underline{3 \text{ yd } 1 \text{ ft } 8 \text{ in}} \\
5 \text{ yd } 3 \text{ ft } 14 \text{ in} = 6 \text{ yd } 1 \text{ ft } 2 \text{ in}
\end{array}
$$

41. $(3 \text{ ft } 8 \text{ in}) + (12 \text{ ft } 10 \text{ in}) + (20 \text{ ft } 8 \text{ in}) + (7 \text{ ft } 4 \text{ in}) = 42 \text{ ft } 30 \text{ in} = 44 \text{ ft } 6 \text{ in}$

43. $7(298 \text{ gm}) = 2086 \text{ gm}$

45. $\dfrac{12° + 15° + 12° + 11° + 10°}{5} = \dfrac{60°}{5} = 12° \text{ C}$

47. 2(20 mg)(3)(7) = 840 mg

49. (50 m)(8 strips) = 400 m

51. The dimensions of the base of the fountain are 4 ft × 4 ft

53. The patio is 16 ft long and 12 ft wide.

55. It is not possible to add different units of measure together.

57. group activity

59. 8 + 56 + 129 + 35 + 604 = 832

61. $5(850 − 68) = $5(782) = $3910

Section 2.2
Perimeter

1. 5 in. + 6 in. + 8 in. = 19 in.

3. 4(7 m) = 28 m

5. 2(12 mm) + 2(5 mm) = 24 mm + 10 mm = 34 mm

7. 2(2 km) + 2(11 km) = 4 km + 22 km = 26 km

9. 16 mm + 27 mm + 40 mm = 83 mm

11. 2(45 m) + 2(35 m) = 160 m

13. 2(15 ft) + 2(30 ft) = 30 ft + 60 ft = 90 ft

15. 2(36 cm) + 2(14 cm) = 72 cm + 28 cm = 100 cm

17. $6(13 \text{ yd}) = 78 \text{ yd}$

19. $19 \text{ m} + 30 \text{ m} + 49 \text{ m} + 19 \text{ m} + 16 \text{ m} + 35 \text{ m} = 168 \text{ m}$

21. $16 \text{ cm} + 5 \text{ cm} + 7 \text{ cm} + 16 \text{ cm} + 2 \text{ cm} + 2 \text{ cm} + 6 \text{ cm} = 54 \text{ cm}$

23. $60 \text{ mm} + 14 \text{ mm} + 21 \text{ mm} + 25 \text{ mm} + 25 \text{ mm} + 48 \text{ mm} + 75 \text{ mm} + 40 \text{ mm} + 10 \text{ mm} + 30 \text{ mm} = 348 \text{ mm}$

25. $4(2)(20 \text{ in}) = 160 \text{ in} = 13 \text{ ft } 4 \text{ in}$

27. $2(15 \text{ ft} + 12 \text{ ft}) = 2(27 \text{ ft}) = 54 \text{ ft};$ $2 \text{ min/ft}(54 \text{ ft}) = 108 \text{ min} = 1 \text{ hr } 48 \text{ min}$

29. The dimensions are 20 inches by 26 inches

31. $12 \text{ ft} + 45 \text{ ft} + 24 \text{ ft} + 12 \text{ ft} + 24 \text{ ft} = 117 \text{ ft}$

33. $2(120 \text{ yd} + 53 \text{ yd}) = 2(173 \text{ yd}) = 346 \text{ yd};$ $346 \text{ yd/lap}(56 \text{ laps}) = 19,376 \text{ yd}$

35. The shape of the patio is an octogon.

37. $4(4 \text{ ft}) = 16 \text{ ft}$

39. $10 \text{ ft} + 3 \text{ ft} + 4 \text{ ft} + 2 \text{ ft} + 8 \text{ ft} + 2 \text{ ft} + 4 \text{ ft} + 3 \text{ ft} = 36 \text{ ft}$

41. A rectangle is always a parallelogram because it has two pair of parallel sides.

43. $2(5 \text{ ft})(4) + 5(6 \text{ ft}) - 4(2 \text{ ft } 6 \text{ in}) = 40 \text{ ft} + 30 \text{ ft} - 8 \text{ ft } 24 \text{ in} = 70 \text{ ft} - 10 \text{ ft} = 60 \text{ ft}$
    $60 \text{ ft}(\$3/\text{ft}) = \$180;$ $\$180 + 4(\$15) = \$180 + \$60 = \$240$

45. nine hundred thousand, fifty = 900,050

47. 32,571,600 rounded to the nearest ten thousand is 32,570,000

49. $733(348) = 255,084$

Section 2.3
Area

1. $(4\,\text{km})^2 = 16\,\text{km}^2$

3. $\dfrac{1}{2}(5\,\text{yd})(8\,\text{yd}) = 20\,\text{yd}^2$

5. $5\,\text{yd}(12\,\text{yd}) = 60$ sq. yd.

7. $\dfrac{1}{2}(4\,\text{m})(10\,\text{m} + 16\,\text{m}) = 2\,\text{m}(26\,\text{m}) = 52\,\text{m}^2$

9. $12\,\text{km}(11\,\text{km}) = 132$ sq km

11. $\dfrac{1}{2}(28\,\text{yd})(17\,\text{yd}) = 238\,\text{yd}^2$

13. $\dfrac{1}{2}(36\,\text{cm})(22\,\text{cm}) = 396\,\text{cm}^2$

15. $15\,\text{in}(9\,\text{in}) = 135$ sq. in.

17. $\dfrac{1}{2}(18\,\text{m})(25\,\text{m} + 43\,\text{m}) = (9\,\text{m})(68\,\text{m}) = 612\,\text{m}^2$

19. $(12\,\text{ft})(8\,\text{ft}) + (5\,\text{ft})(1\,\text{ft}) = 96\,\text{ft}^2 + 5\,\text{ft}^2 = 101\,\text{ft}^2$

21. $(50\,\text{m})(140\,\text{m}) - 2(15\,\text{m}^2) = 7000\,\text{m}^2 - 2(225\,\text{m}^2) = 7000\,\text{m}^2 - 450\,\text{m}^2 = 6550\,\text{m}^2$

23. $(65\,\text{in})(85\,\text{in}) - \dfrac{1}{2}(20\,\text{in})(65\,\text{in}) + \dfrac{1}{2}(24\,\text{in})(65\,\text{in}) -$
    $(5525\,\text{in}^2) - (650\,\text{in}^2) + (780\,\text{in}^2) = 5655\,\text{in}^2$

25. $9\,\text{ft}^2 = 1\,\text{yd}^2$

27. Two gallons of stain will cover $2(250\,\text{ft}^2) = 500\,\text{ft}^2$.
    The area of the side of the house with no windows is $(35\,\text{ft})(22\,\text{ft}) = 770\,\text{ft}^2$.
    Therefore, two gallons of stain is not enough to cover the side.

29. The area of the lawn is $(8\,\text{m})(30\,\text{m}) = 240\,\text{m}^2$. Since one ounce of weed killer treats one square meter of lawn, then it requires 240 oz. to treat 240 sq. m.

31. $2(3\,\text{ft})(6\,\text{ft}) = 36$ sq. ft.

33. $\dfrac{1}{2}(9\,\text{ft})(36\,\text{ft}) = 162\,\text{ft}^2$

35. $2\left(\dfrac{1}{2}\right)(18\,\text{in})(50\,\text{in} + 24\,\text{in}) = 18(74\,\text{in}) = 1332\,\text{in}^2$

37. The area is:
$$\frac{1}{2}(10\,\text{ft})(10\,\text{ft} + 16\,\text{ft}) + \frac{1}{2}(16\,\text{ft})(30\,\text{ft} + 20\,\text{ft}) - (15\,\text{ft})(8\,\text{ft}) =$$
$130\,\text{ft}^2 + 400\,\text{ft}^2 + 120\,\text{ft}^2 = 410\,\text{ft}^2$
One bag covers 70 sq. ft; therefore: $410\,\text{ft}^2 \div 70\,\text{ft}^2 \approx 6$ bags are needed.

39. The total number of bricks required to cover the patio and the walkways is:
7(322 sq.ft.) = 2254 bricks

41. answers vary

43. The area of the kitchen floor is 9 ft(10 ft) = 90 sq.ft.
The number of black six-inch tiles that are needed is:
2(10)(2) + 2(8)(2) = 40 + 32 = 72 6-inch tiles
The number of white six-inch tiles that are needed is:
2(9)(2) + 2(7)(2) = 36 + 28 = 64 6-inch tiles
The number of black one-foot tiles that are needed is:
2(8) + 2(5) = 16 + 10 = 26 plus 2(4) + 2(1) = 8 + 2 = 10 ⇒ 36 1-foot tiles
The number of white one-foot tiles that are needed is:
2(6) + 2(3) = 12 + 6 = 18 plus 2(1) = 2 ⇒ 20 1-foot tiles

45. 30 ft(24 ft) = 720 sq ft   22 ft(18 ft) = 396 sq. ft.
The total area to be carpeted is 720 sq.ft. + 396 sq.ft. = 1116 sq. ft
or $1116\,\text{sq.ft.}\left(\dfrac{1\,\text{sq. yd.}}{9\,\text{sq. ft.}}\right) = 124\,\text{sq.yd.}$
The cost to have the carpet installed is 124 sq.yd.($27.00/sq.yd.) = $3348

47. group activity

49. $295,850 \div 97 = 3050$.

51. $4 + 2($6) + 1($4) + 3($3)
$4 + $12 + $4 + $9
$29 is the cost for the family to attend the exhibition

Section 2.4
Volume

1. $20 \text{ m}(5 \text{ m})(10 \text{ m}) = 1000 \text{ m}^3$

3. $5 \text{ ft}(5 \text{ ft})(10 \text{ ft}) = 250 \text{ ft}^3$

5. $10 \text{ cm}(12 \text{ cm}^2) = 120 \text{ cm}^3$

7. $52 \text{ cm}(35 \text{ cm})(12 \text{ cm}) = 21,840 \text{ cm}^3 = 21,840 \text{ ml}$

9. $4 \text{ ft}(12 \text{ ft}^2) = 48 \text{ ft}^3$

11. $15 \text{ in}(6 \text{ in})(34 \text{ in}) = 3060 \text{ in}^3$

13. $355 \text{ cm}^2(122 \text{ cm}) = 43,310 \text{ cm}^3$

15. $(245 \text{ in}^2)(24 \text{ in}) = 5880 \text{ in}^3$

17. $5184 \text{ in}^3 \div 1728 \text{ in}^3/\text{ft}^3 = 3 \text{ ft}^3$

19. $\dfrac{270 \text{ ft}^3}{1} \cdot \dfrac{1 \text{ yd}^3}{27 \text{ ft}^3} = 10 \text{ yd}^3$

21. $(3 \text{ ft})(4 \text{ in})(54 \text{ ft}) = \left(3 \text{ ft} \cdot \dfrac{12 \text{ in}}{\text{ft}}\right)(4 \text{ in})\left(54 \text{ ft} \cdot \dfrac{12 \text{ in}}{\text{ft}}\right) = (36 \text{ in})(4 \text{ in})(648 \text{ in}) = 93,312 \text{ in}^3$

    $93,312 \text{ in}^3 \cdot \dfrac{\text{yd}^3}{(36 \text{ in})^3} = 93,312 \text{ in}^3 \cdot \dfrac{\text{yd}^3}{46,656 \text{ in}^3} = 2 \text{ yd}^3$

23. $\dfrac{1}{2}(6 \text{ in})(12 \text{ in} + 8 \text{ in})(20 \text{ in}) = 3 \text{ in}(20 \text{ in})(20 \text{ in}) = 1200 \text{ in}^3$

25. $3 \text{ yd}(16 \text{ yd})(10 \text{ yd}) + 2\left(\dfrac{1}{2}\right)(3 \text{ yd})(12 \text{ yd} + 16 \text{ yd})(10 \text{ yd}) = 480 \text{ yd}^3 + 840 \text{ yd}^3 = 1320 \text{ yd}^3$

27. The volume is:

    $(15 \text{ cm})(72 \text{ cm})(54 \text{ cm}) + (110 \text{ cm})(72 \text{ cm})(15 \text{ cm})$

    $58,320 \text{ cm}^3 + 118,800 \text{ cm}^3$

    $177,120 \text{ cm}^3$

29. $(24\text{ ft})(36\text{ ft})(7\text{ ft}) = 6048\text{ ft}^3$ $\qquad\qquad$ $6048\text{ ft}^3 \div 378\text{ ft}^3 = 16$ truckloads

31. The number of bricks needed to cover the patio and the pathways is 2254.

$$\frac{2254\text{ bricks}}{1} \cdot \frac{\text{ft}^3}{112\text{ bricks}} \approx 20\text{ ft}^3$$

33. The total perimeter of the slab is:

$(3 + 7 + 7 + 40 + 12 + 8 + 8 + 3 + 4 + 2 + 4 + 6 + 13 + 4 + 6 + 16 + 3 + 4 + 2$
$+ 4 + 6 + 5 + 4 + 3 + 10 + 3 + 4 + 2 + 8 + 2 + 16)\text{ ft} = 219\text{ ft}$

35. answers vary

37. The volume of the pool is:

$$10\text{ ft}(15\text{ ft})(3\text{ ft}) + \frac{1}{2}(20\text{ ft})(10\text{ ft} + 3\text{ ft})(15\text{ ft})$$

$$450\text{ ft}^3 + (10\text{ ft})(13\text{ ft})(15\text{ ft})$$

$$450\text{ ft}^3 + 1950\text{ ft}^3$$

$$2400\text{ ft}^3$$

$$2400\text{ ft}^3\left(\frac{\text{yd}}{3\text{ ft}}\right)^3 = 2400\text{ ft}^3\left(\frac{\text{yd}^3}{27\text{ ft}^3}\right) = 88.89\text{ yd}^3$$

$88.89\text{ yd}^3 \div 14\text{ yd}^3 = 6.35$ loads $\Rightarrow$ 7 loads of dirt hauled away

39. group activity

41. $4^2 + 5^2 = 16 + 25 = 41$

43. $9^2 - 4^2 = 81 - 16 = 65$

Chapter 2
True-False Concept Review

1. False – Metric measurements are the most commonly used in the world.

2. True

3. False – A liter is a measure of capacity.

4. True

5. False – Volume is the measure of the inside of a solid such as a box or a can.

6. False – The volume of a cube is $V = s^3$

7. True

8. True

9. True

10. True

11. True

12. False – A parallelogram has four sides.

13. False – Area can be thought of as the number of squares in an object.
    Volume can be thought of as the number of cubes in an object.

14. True

15. True

16. True

17. False – Volume can be measured in cubic units and units such as liters, quarts, etc.

18. True

19. False – $1\,\text{ft}^2 = 144\,\text{in}^2$

20. False – The prefix "kilo" means 1000.

Chapter 2
Review

1. $(6 \text{ ft})5 = 30 \text{ ft}$ 3. $6(4 \text{ ft}) = 24 \text{ ft}$ 5. $25(190 \, l) = 4750 \, l$

7. $2 \text{ g} - 350 \text{ mg} = 2000 \text{ mg} - 350 \text{ mg} = 1650 \text{ mg}$

9. $\quad 45 \text{ gal 2 qt} = 44 \text{ gal 6 qt}$
    $\quad \underline{- \, 18 \text{ gal 3 qt} = 18 \text{ gal 3 qt}}$
    $\qquad\qquad\quad 26 \text{ gal 3 qt}$

11. $3 \text{ ft} + 4 \text{ ft} + 5 \text{ ft} = 12 \text{ ft}$

13. $8(4 \text{ m}) = 32 \text{ m}$

15. $2(1500 \text{ ft} + 250 \text{ ft}) = 2(1750 \text{ ft}) = 3500 \text{ ft};$ $3500 \text{ ft}(\$5/\text{ft}) = \$17,500$

17. $(21 \text{ ft})(9 \text{ ft}) = 819 \text{ sq.ft.}$ 19. $(3 \text{ ft})(18 \text{ in}) = (36 \text{ in})(18 \text{ in}) = 648 \text{ in}^2$

21. $(12 \text{ in})(8 \text{ in})(3 \text{ in}) = 288 \text{ in}^3$ 23. $40 \text{ ft}(8 \text{ ft})(6 \text{ ft}) = 1920 \text{ ft}^3$

25. $(7 \text{ in})^3 = 343 \text{ in}^3$

Chapter 2
Test

1. $7 \text{ m} + 454 \text{ mm} = 7000 \text{ mm} + 454 \text{ mm} = 7454 \text{ mm}$

2. $4(34 \text{ cm}) = 136 \text{ cm}$

3. $4 \text{ in}(18 \text{ in})(24 \text{ in}) = 1728 \text{ cu. in.}$

4. $8 \text{ ft}(5 \text{ ft}) + 4 \text{ ft}(10 \text{ ft}) = 40 \text{ ft}^2 + 40 \text{ ft}^2 = 80 \text{ ft}^2$

5. $135 \text{ lb} \div 5 = 27 \text{ lb}$

6. $6 \text{ ft} + 4 \text{ ft} + 6 \text{ ft} + 3 \text{ ft} + 3 \text{ ft} = 22 \text{ ft}$

7. $\frac{1}{2}(15\,\text{in})(4\,\text{in}) = 30\,\text{in}^2$

8.
$$\begin{array}{ll} 5\ \text{gal}\ 2\ \text{qt} & = 4\ \text{gal}\ 6\ \text{qt}\ 2\ \text{pt} \\ -\ \underline{3\ \text{gal}\ 2\ \text{qt}\ 1\ \text{pt}} & = \underline{3\ \text{gal}\ 2\ \text{qt}\ 1\ \text{pt}} \\ & \quad\ 1\ \text{gal}\ 3\ \text{qt}\ 1\ \text{pt} \end{array}$$

9. answers vary

10. $2\,\text{ft}^2 \times \dfrac{144\,\text{in}^2}{\text{ft}^2} = 288\,\text{in}^2$

11. The amount of weather stripping needed is:
$$2(4\,\text{ft} + 5\,\text{ft}) + 2(2)(2\,\text{ft} + 5\,\text{ft})$$
$$2(9\,\text{ft}) + 4(7\,\text{ft})$$
$$18\,\text{ft} + 28\,\text{ft}$$
$$46\,\text{ft}$$

12. $\frac{1}{2}(6\,\text{m})(12\,\text{m} + 10\,\text{m})(4\,\text{m}) = 3\,\text{m}(22\,\text{m})(4\,\text{m}) = 66\,\text{m}^2(4\,\text{m}) = 264\,\text{m}^3$

13. $(8\,\text{cm})(6\,\text{cm}) = 48\ \text{cm}^2$

14. $(5\,\text{ft})(4\,\text{ft}^2) = 20\ \text{ft}^3$

15. $684\,\text{g} \div 6 = 114\,\text{g}$

16. $3(6\,\text{ft}) + 4(12\,\text{ft}) = 18\,\text{ft} + 48\,\text{ft} = 66\,\text{ft};$ $66\,\text{ft}(\$2) = \$132$

17. answers vary
    examples – English system: cups, pints, quarts, gallons, etc.
    examples – metric system: liters, milliliters, kiloliters, etc.

18. $36\,\text{mm}(90\,\text{mm}) - \frac{1}{2}(36\,\text{mm})(28\,\text{mm}) = 3240\ \text{mm}^2 - 504\ \text{mm}^2 = 2736\ \text{mm}^2$

19. $2(100\,\text{yd} + 60\,\text{yd}) = 2(160\,\text{yd}) = 320\,\text{yd};$ $320\,\text{yd}(6) = 1920\,\text{yd}$

20. Area measures the interior space of a two-dimensional object.
    Volume measures the interior space of a three-dimensional object.

CHAPTER THREE
INTRODUCTION TO ALGEBRA

Section 3.1
The Language of Algebra

1. Twelve more than a number: $12 + n$

3. The product of seven and a number: $7n$

5. The difference of twenty and a number: $20 - n$

7. The quotient of a number and five: $\dfrac{n}{5}$

9. Eighteen more than twice a number: $2n + 18$

11. Fifteen less than the quotient of a number and two: $\dfrac{n}{2} - 15$

13. The difference of eight times a number and twice the same number: $8n - 2n$

15. The product of two numbers decreased by fifteen: $xy - 15$

17. The quotient of two numbers increased by 20: $\dfrac{x}{y} + 20$

19. The product of fifteen and two less than a number: $15(n - 2)$

21. $32 + a = 32 + 6 = 38$

23. $5a = 5(6) = 30$

25. $\dfrac{72}{a} = \dfrac{72}{6} = 12$

27. $a + 98 = 6 + 98 = 104$

29. $a^2 = 6^2 = 36$

31. $17a - 3 = 17(6) - 3 = 102 - 3 = 99$

33. $a^2 + 6 = 6^2 + 6 = 36 + 6 = 42$

35. $xy = 56(14) = 784$

37. $2x - 5y = 2(56) - 5(14) = 112 - 70 = 42$

39. $\dfrac{x}{y+14} = \dfrac{56}{14+14} = \dfrac{56}{28} = 2$

41. Replacing x with 17 in $x + 4$ is called <u>substitution.</u>

43. $2(92) - 100 = 184 - 100 = 84$

45. $\$18n$, $\$18(4) = \$72$, $\$18(7) = \$126$, $\$18(12) = \$216$

47. $\$16n + \$3$ $\$16(2) + \$3 = \$32 + \$3 = \$35$
    $\$16(5) + \$3 = \$80 + \$3 = \$83$ $\$16(8) + \$3 = \$128 + \$3 = \$131$

49. $\$11,250x$ $\$11,250(5) = \$56,250$
    $\$11,250(11) = \$123,750$ $\$11,250(14) = \$157,500$

51. $(a + b + c)h$

53. a. $\dfrac{60 + 70 + 85 + 52 + 58 + 58 + 43 + 56 + 46 + 52}{10} = \dfrac{580}{10} = 58$

    b. $\dfrac{a + b + c + d + e + f + g + h + i + j}{10} =$
    average number of accidental drownings over a ten-year period

55. $B = \dfrac{x}{100}$ 57. answers vary   59. answers vary

61. Evaluate:
    $a^2 + 5a - 3b - c$
    $(20)^2 + 5(20) - 3(13) - 5$
    $400 + 100 - 39 - 5$
    456

63. Evaluate: $\dfrac{6a + 10b}{25} = \dfrac{6(20) + 10(13)}{25} = \dfrac{120 + 130}{25} = \dfrac{250}{25} = 10$

65. Evaluate:

$$a^3 - b^2 - c^2 - x^2$$
$$20^3 - 13^2 - 5^2 - 7^2$$
$$8000 - 169 - 25 - 49$$
$$7757$$

67. answers vary

69. $34 + 86 + 38 + 65 = 223$

71. $8 + 56 + 129 + 35 + 604 = 832$

Section 3.2
Equations and Formulas

1. 5 is a solution
$$x + 16 = 21$$
$$5 + 16 = 21$$
$$21 = 21$$

3. 12 is a solution
$$22 - y = 10$$
$$22 - 12 = 10$$
$$10 = 10$$

5. 9 is not a solution
$$w - 10 = 1$$
$$9 - 10 = 1$$
$$-1 \neq 1$$

7. 7 is a solution
$$10 + 2a = 24$$
$$10 + 2(7) = 24$$
$$10 + 14 = 24$$
$$24 = 24$$

9. 3 is a solution
$$\frac{45}{x} = 15$$
$$\frac{45}{3} = 15$$
$$15 = 15$$

11. 23 is a solution
$$3w - 15 = 54$$
$$3(23) - 15 = 54$$
$$69 - 15 = 54$$
$$54 = 54$$

13. 13 is not a solution

$$55 - 2t = 33$$

$$55 - 2(13) = 33$$

$$55 - 26 = 33$$

$$29 \neq 33$$

15. 42 is not a solution

$$764 - 2m = 670$$

$$764 - 2(42) = 670$$

$$764 - 84 = 670$$

$$680 \neq 670$$

17. 22 is a solution

$$\frac{286}{x} = 13$$

$$\frac{286}{22} = 13$$

$$13 = 13$$

19. 18 is not a solution

$$\frac{5y}{6} + 4 = 18$$

$$\frac{5(18)}{6} + 4 = 18$$

$$\frac{90}{6} + 4 = 18$$

$$15 + 4 = 18$$

$$19 \neq 18$$

21. $A = \dfrac{bh}{2} = \dfrac{2\,\text{ft} \cdot 3\,\text{ft}}{2} = 3\,\text{ft}^2$

23. $D = rt = 55\,\text{mph}(2\,\text{hr}) = 110\,\text{mi}$

25. $P = a + b + c = 12\,\text{in} + 4\,\text{in} + 9\,\text{in} = 25\,\text{in}$

27. $h = \dfrac{2A}{b} = \dfrac{2(14\,\text{cm}^2)}{14\,\text{cm}} = 2\,\text{cm}$

29. $V = s^3 = (16\,\text{m})^3 = 4096\,\text{m}^3$

31. $s = \dfrac{P}{4} = \dfrac{388\,\text{ft}}{4} = 97\,\text{ft}$

33. $V = lwh = (12\,\text{in})(12\,\text{in})(5\,\text{in}) = 720\,\text{in}^3$

35. $P = 2l + 2w = 2(203\,\text{ft}) + 2(76\,\text{ft}) = 406\,\text{ft} + 152\,\text{ft} = 558\,\text{ft}$

37. Solve:

$$x + 8 = 15$$

$$x + 8 - 8 = 15 - 8$$

$$x = 7$$

39. Solve:

$$y - 3 = 12$$

$$y - 3 + 3 = 12 + 3$$

$$y = 15$$

41. Solve:

$$\frac{w}{5} = 7$$

$$\frac{5 \cdot w}{5} = 7 \cdot 5$$

$$w = 35$$

43. Solve:

$$5m = 55$$

$$\frac{5m}{5} = \frac{55}{5}$$

$$m = 11$$

45. Solve:
$$2x + 3 = 17$$
$$2x + 3 - 3 = 17 - 3$$
$$2x = 14$$
$$\frac{2x}{2} = \frac{14}{2}$$
$$x = 7$$

47. Solve:
$$2y - 5 = 19$$
$$2y - 5 + 5 = 19 + 5$$
$$2y = 24$$
$$\frac{2y}{2} = \frac{24}{2}$$
$$y = 12$$

49. Solve:
$$54 + 2w = 70$$
$$54 - 54 + 2w = 70 - 54$$
$$2w = 16$$
$$\frac{2w}{2} = \frac{16}{2}$$
$$w = 8$$

51. Solve:
$$85 - 5a = 30$$
$$85 - 85 - 5a = 30 - 85$$
$$-5a = -55$$
$$\frac{-5a}{-5} = \frac{-55}{-5}$$
$$a = 11$$

53. The statement "$x + 5 = x$" is a <u>false</u> equation.

55. Every equation has <u>two</u> members.

57. $S = \dfrac{n(a+l)}{2} = \dfrac{25(1+25)}{2} = \dfrac{25(26)}{2} = 25(13) = 325$

59. $S = \dfrac{n(a+l)}{2} = \dfrac{100(2+200)}{2} = \dfrac{100(202)}{2} = 100(101) = 10,100$

61. 18 ft(25 ft) = 450 sq. ft.

63. $T = D + pm = \$85 + \$41(12) = \$85 + \$492 = \$577$

65. Specific Gravity $= \dfrac{\text{weight in air}}{\text{weight in air - weight in water}} = \dfrac{400 \text{ gm}}{400 \text{ gm} - 320 \text{ gm}} = \dfrac{400 \text{ gm}}{80 \text{ gm}} = 5$

67. $\text{IQ} = \dfrac{100(\text{MA})}{\text{CA}} = \dfrac{100(32)}{25} = \dfrac{3200}{25} = 128$

69. $E = 3I = 3(\$2300) = \$6900$; $\qquad E = 6I = 6(\$2300) = \$13,800$

71.

No. of Guests	Hors d'oeuvres	Bartenders	Cocktail Servers
x	$H = 18x - 24x$	$B = \dfrac{x}{100}$	$C = \dfrac{x}{50}$
800	$H = 18(800) - 24(800)$ $= 14{,}400 - 19{,}200$	$B = \dfrac{800}{100} = 8$	$C = \dfrac{800}{50} = 16$
900	$H = 18(900) - 24(900)$ $= 16{,}200 - 21{,}600$	$B = \dfrac{900}{100} = 9$	$C = \dfrac{900}{50} = 18$
1000	$H = 18(1000) - 24(1000)$ $= 18{,}000 - 24{,}000$	$B = \dfrac{1000}{100} = 10$	$C = \dfrac{1000}{50} = 20$
1100	$H = 18(1100) - 24(1100)$ $= 19{,}800 - 26{,}400$	$B = \dfrac{1100}{100} = 11$	$C = \dfrac{1100}{50} = 22$

73. To find the area of a rectangle, multiply its length by its width. **Answers vary.**

75. 16 is a solution

$$x^2 - 7x = 144$$
$$16^2 - 7(16) = 144$$
$$256 - 112 = 144$$
$$144 = 144$$

77. 13 is not a solution

$$2x^2 - 10x = 221$$
$$2(13)^2 - 10(13) = 221$$
$$2(169) - 130 = 221$$
$$338 - 130 = 221$$
$$208 \neq 221$$

79. If $900\text{ ft} = 2l + w$, then $900 - 2l = w$. The area of the corral is: $A = l(900 - 2l)$.

81. $7844 + 609 + 2608 + 87 = 11{,}148$

83. $28\text{ cm} + 24\text{ cm} + 37\text{ cm} = 89\text{ cm}$

Section 3.3
Adding and Subtracting Polynomials (Whole Numbers)

1. The coefficient of $75t$ is 75.

3. The like terms are: $3xy$ and $8xy$.

5. The like terms are: $14m^2n$ and $39m^2n$.

7. The coefficient of ab^2 is 1.

9. The like terms are: $17pq$, pq and $13pq$.

11. $10a + 11a = 21a$

13. $14c + 13c = 27c$

15. $31x + 8x + x = 40x$

17. $15y - 9y = 6y$

19. $36w + 14w = 50w$

21. $14x + 25x + 15x + 22x = 76x$

23. $48y + 33y - 13y - 9y = 59y$

25. $103s + 219s - 78s = 244s$

27. $38w - 19w + 56w - w = 74w$

29. $(2x + 5y) + (6x + 2y) = 8x + 7y$

31. $7r + 2t + 19r + 7t = 26r + 9t$

33. $17x^2 + 5x + 2x^2 + 18x = 19x^2 + 23x$

35. $m^2 + s + 2s + 3m^2 = 4m^2 + 3s$

37. $10x - 7x + 10y - 2y = 3x + 8y$

39. Simplify:
$$(49xy + 39yz) - (31xy + 17yz)$$
$$49xy + 39yz - 31xy - 17yz$$
$$18xy + 22yz$$

41. Simplify: $(55x^2 + 16x) + (39x^2 + 6x) = 94x^2 + 22x$

43. Simplify:
$$(43x^2 + 17x + 75) - (15x^2 + 9x + 19)$$
$$43x^2 + 17x + 75 - 15x^2 - 9x - 19$$
$$28x^2 + 8x + 56$$

45. Simplify: $(12xy + 17yz + 19wz) + (11xy + 26yz + 18wz) = 23xy + 43yz + 37wz$

47. Simplify: $(15x^2 + 19 + 22x) + (2x^2 + 12x + 18) = 17x^2 + 34x + 37$

49. Simplify: $(15pq + 25qs) - (7pq + 12qs) = 15pq + 25qs - 7pq - 12qs = 8pq + 13qs$

51. Simplify:
$(72abc + 83ab + 95c) - (68abc + 77ab + 89c)$
$\quad 72abc + 83ab + 95c - 68abc - 77ab - 89c$
$\qquad\qquad 4abc + 6ab + 6c$

53. Only <u>like</u> terms can be combined.

55. $45x^2 + 2x = 47x^3$ is not correct because unlike terms cannot be added together.
The above equation will be correct if it is changed to $45x^2 + 2x^2 = 47x^2$

57. $P = 2(w + 5) + 2(3w + 1) = 2w + 10 + 6w + 2 = 8w + 12$

59. $T = i + 2i = 3i = 3(\$12{,}000) = \$36{,}000$

61. $d = \dfrac{\$15}{100\,\text{mi}}\,m - \dfrac{\$11}{100\,\text{mi}}\,m = \dfrac{\$4}{100\,\text{mi}}\,m = \dfrac{\$4}{100\,\text{mi}}(800\,\text{mi}) = \$32$

63. $L = y + y + 2y + 2y + 2y = 8y = 8(10\,\text{ft}) = 80\,\text{ft}$

65. $4(x + 3)\,\text{ft} + (x + 2)\,\text{ft} + (x + 3 + 3 + 4 + 4)\,\text{ft} = 4x\,\text{ft} + 12\,\text{ft} + 2x\,\text{ft} + 16\,\text{ft} = (6x + 28)\,\text{ft}$

67. $1351P + 309Q$

69. $14B + 6C$; $\quad 14\left(\dfrac{x}{100}\right) + 6\left(\dfrac{x}{50}\right) = \dfrac{13x}{50}$

71. $R = 6x + 14\left(\dfrac{x}{100}\right) + 6\left(\dfrac{x}{50}\right) = \dfrac{13x}{50}$

73. $(24xy + 4x^2) + (2x^2 + 17xy) = 6x^2 + 41xy$ (add like terms)

75. $(75a + 52b + 111c) + (38a - 47b) = 113a + 5b + 111c$

77. $(28z + 12z) + (8z - 6z) = 40z + 2z = 42z$

79. $(28z + 12z) - (8z - 6z) = 40z - 2z = 38z$

81. answers vary

83. $107 > 99$ is true

85. Total area:
$2\,\text{m}(2\,\text{m})(3\,\text{m}) + 2\,\text{m}(2\,\text{m})(3\,\text{m}) + 2\,\text{m}(2\,\text{m})(2\,\text{m}) = 12\,\text{m}^2 + 12\,\text{m}^2 + 8\,\text{m}^2 = 32\,\text{m}^2$

Section 3.4
Multiplying and Dividing Polynomials (Whole Numbers)

1. $4(12w) = 48w$

3. $7(9a^2b) = 63a^2b$

5. $\dfrac{48y}{6} = 8y$

7. $\dfrac{72w}{6} = 12w$

9. $10(32bc) = 320bc$

11. $\dfrac{39x}{3} = 13x$

13. $(14xy)(12y) = 168xy^2$

15. $(21st)(19st) = 399s^2t^2$

17. $\dfrac{312abc}{24b} = 13ac$

19. $\dfrac{152d^2ef}{19ef} = 8d^2$

21. $\dfrac{16ax}{4x} + \dfrac{12bx}{4x} + \dfrac{20cx}{4x} = 4a + 3b + 5c$

23. $5z(11z + 2) = 55z^2 + 10z$

25. $8xy(7yz - 2xz) = 56xy^2z - 16x^2yz$

27. $2xy(15xz - 22yz + 25xy) = 30x^2yz - 44xy^2z + 50x^2y^2$

29. $\dfrac{16by + 12cy + 30dy}{4y} = 4b + 3c + 5d$

31. $\dfrac{44xyz^2 + 55x^2yz + 88xy^2z}{11xy} = 4z^2 + 5xz + 8yz$

33. $21abc(4a + 3b - 5c) = 84a^2bc + 63ab^2c - 105abc^2$

35. $(y + 6)(y + 3) = y^2 + 9y + 18$

37. $(x + 15)(x + 2) = x^2 + 17x + 30$

39. $\dfrac{336x^2 - 352xw}{16x} = 21x - 22w$

41. $\dfrac{324a^2bc - 234ab^2c + 162abc^2}{18abc} = 18a - 13b + 9c$

43. $19a(4a)(23a) = 1748a^3$

45. $(3x+7)(2x+5) = 6x^2 + 29x + 35$

47. $(6x+1)(4x+9) = 24x^2 + 58x + 9$

49. $17x(5x) = (17 \cdot 5)(x \cdot x) = 85x^2$

51. The error in the statement $17x(5y+2) = 85xy + 2$ is that 17x has not been multiplied by the 2. The correct statement is $17x(5y+2) = 85xy + 34x$.

53. $A = w(2w) = 2w^2 = 2(11\,\text{ft})^2 = 2(121\,\text{ft}^2) = 242\,\text{ft}^2$

55. $V = w(2w)(3w) = 6w^3 = 6(2\,\text{ft})^3 = 6(8\,\text{ft}^3) = 48\,\text{ft}^3$

57. $A = \dfrac{4b(b)}{2} = 2b^2 = 2(16\,\text{in}^2)^2 = 2(256\,\text{in}^2) = 512\,\text{in}^2$

59. $W = \dfrac{18C}{12} + 16 = \dfrac{18C}{12} + 16 \cdot \dfrac{12}{12} = \dfrac{18C + 192}{12}$

61. RC represents the number of total chairs.

63. Total area of the hall:

C = Chairs/Row	R = # of Rows	Total Seating Capacity	Total Area of Hall
25	10	250	408,312 sq.in. or 2835.5 sq.ft.
20	13	260	420,624 sq.in. or 2921 sq.ft.
16	16	256	426,240 sq.in. or 2960 sq.ft.
10	25	250	470,952 sq.in. or 3270.5 sq.ft.

65. $(2g^2+3)(5g^2+5) = 10g^4 + 10g^2 + 15g^2 + 15 = 10g^4 + 25g^2 + 15$

67. $\dfrac{48x^2y^2}{4xy} = 12xy$

69. $V = 4x(3x)(2x+1) = 12x^2(2x+1) = 24x^3 + 12x^2$

71. $V = 3r^2(3r+4) = 9r^3 + 12r^2$

73. $3412 - 1987 = 1425$

75. $42 + 3n$

77. $V = 15\,\text{in}(12\,\text{in})(7\,\text{in}) = 1260\,\text{in}^3$

Section 3.5
Solving Equations of the Form $x + a = b$ or $x - a = b$

1. Solve:
$$a + 11 = 38$$
$$a + 11 - 11 = 38 - 11$$
$$a = 27$$

3. Solve:
$$c - 5 = 27$$
$$c - 5 + 5 = 27 + 5$$
$$c = 32$$

5. Solve:
$$r + 12 = 20$$
$$r + 12 - 12 = 20 - 12$$
$$r = 8$$

7. Solve:
$$t - 20 = 6$$
$$t - 20 + 20 = 6 + 20$$
$$t = 26$$

9. Solve:
$$y + 16 = 16$$
$$y + 16 - 16 = 16 - 16$$
$$y = 0$$

11. Solve:
$$m - 13 = 13$$
$$m - 13 + 13 = 13 + 13$$
$$m = 26$$

13. Solve:
$$a + 22 = 36$$
$$a + 22 - 22 = 36 - 22$$
$$a = 14$$

15. Solve:
$$c - 42 = 60$$
$$c - 42 + 42 = 60 + 42$$
$$c = 102$$

17. Solve:
$$r + 37 = 76$$
$$r + 37 - 37 = 76 - 37$$
$$r = 39$$

19. Solve:
$$t - 25 = 67$$
$$t - 25 + 25 = 67 + 25$$
$$t = 92$$

21. Solve:
$$y + 92 = 120$$
$$y + 92 - 92 = 120 - 92$$
$$y = 28$$

23. Solve:
$$m - 76 = 76$$
$$m - 76 + 76 = 76 + 76$$
$$m = 152$$

25. Solve:
$$21 + y = 53$$
$$21 - 21 + y = 53 - 21$$
$$y = 32$$

27. Solve:
$$65 = t - 15$$
$$65 + 15 = t - 15 + 15$$
$$80 = t$$

29. Solve:
$$134 = 26 + z$$
$$134 - 26 = 26 - 26 + z$$
$$108 = z$$

31. Solve:
$$98 + a = 103$$
$$98 - 98 + a = 103 - 98$$
$$a = 5$$

33. Solve:
$$37 = c - 111$$
$$37 + 111 = c - 111 + 111$$
$$148 = c$$

35. Solve:
$$902 = 15 + u$$
$$902 - 15 = 15 - 15 + u$$
$$887 = u$$

37. Solve:
$$389 + y = 620$$
$$389 - 389 + y = 620 - 389$$
$$y = 231$$

39. Solve:
$$251 = p - 769$$
$$251 + 769 = p - 769 + 769$$
$$1020 = p$$

41. If $59 + w = 78$, then $w = 19$

43. The statement $98 + x = 100$ is true if $x = 2$. The value 98 needs to be subtracted, not added, to both sides of the equation.

45. length of 3rd side:
$$675 = x + 165 + 273$$
$$675 = x + 438$$
$$675 - 438 = x + 438 - 438$$
$$237 \, in = x$$

47. length of 4th side:
$$922 = x + 343 + 133 + 209$$
$$922 = x + 685$$
$$922 - 685 = x + 685 - 685$$
$$237 \, m = x$$

49. a. $m = s + n$, where m = total amount of money budgeted,
s = amount of money spent, and n = amount of money not yet spent

b. Land Acquisition:
$$5,044,999 = 3,463,827 + n$$
$$5,044,999 - 3,463,827 = 3,463,827 - 3,463,827 + n$$
$$1,581,172 = n$$

Open Space:
$$2,555,611 = 2,257,059 + n$$
$$2,555,611 - 2,257,059 = 2,257,059 - 2,257,059 + n$$
$$298,552 = n$$

Pathways Development:
$$3,279,118 = 3,125,675 + n$$
$$3,279,118 - 3,125,675 = 3,125,675 - 3,125,675 + n$$
$$153,443 = n$$

Playfield Improvements:
$$2,367,045 = 1,364,825 + n$$
$$2,367,045 - 1,364,825 = 1,364,825 - 1,364,825 + n$$
$$1,002,220 = n$$

51. The number is:
$$n - 44 = 56$$
$$n - 44 + 44 = 56 + 44$$
$$n = 100$$

53. The Impreza's rating is:
$$x = y + 4$$
$$28 = y + 4$$
$$28 - 4 = y + 4 - 4$$
$$24 \text{ mpg} = y$$

55. The average high temperature in July:
$$x = 26° + y$$
$$85° = 26° + y$$
$$85° - 26° = 26° - 26° + y$$
$$59°F = y$$

57. The number of guests in double rooms is:
$$876 = 568 + d$$
$$308 = d$$

59. $(1046 - 734) - (876 - 616) = 312 - 260 = 52$ more guests in single rooms

61. answers vary   63. answers vary

65. $295{,}850 \div 97 = 3050$ rounded to the nearest thousand is 3000

67. $\dfrac{3n}{71}$

Section 3.6

Solving Equations of the Form $ax = b$ or $\dfrac{x}{a} = b$

1. Solve:
$$5x = 20$$
$$\frac{5x}{5} = \frac{20}{5}$$
$$x = 4$$

3. Solve:
$$7u = 49$$
$$\frac{7u}{7} = \frac{49}{7}$$
$$u = 7$$

5. Solve:
$$8v = 72$$
$$\frac{8v}{8} = \frac{72}{8}$$
$$v = 9$$

7. Solve:
$$2y = 68$$
$$\frac{2y}{2} = \frac{68}{2}$$
$$y = 34$$

9. Solve:
$$\frac{x}{2} = 10$$
$$\frac{x}{2} \cdot 2 = 10 \cdot 2$$
$$x = 20$$

11. Solve:
$$\frac{y}{7} = 9$$
$$\frac{y}{7} \cdot 7 = 9 \cdot 7$$
$$y = 63$$

13. Solve:

$$\frac{y}{12} = 3$$

$$\frac{y}{12} \cdot 12 = 3 \cdot 12$$

$$y = 36$$

15. Solve:

$$\frac{c}{38} = 10$$

$$\frac{c}{38} \cdot 38 = 10 \cdot 38$$

$$c = 380$$

17. Solve:

$$5t = 65$$

$$\frac{5t}{5} = \frac{65}{5}$$

$$t = 13$$

19. Solve:

$$9w = 198$$

$$\frac{9w}{9} = \frac{198}{9}$$

$$w = 22$$

21. Solve:

$$12s = 132$$

$$\frac{12s}{12} = \frac{132}{12}$$

$$s = 11$$

23. Solve:

$$\frac{s}{12} = 132$$

$$\frac{s}{12} \cdot 12 = 132 \cdot 12$$

$$s = 1584$$

25. Solve:

$$13u = 910$$

$$\frac{13u}{13} = \frac{810}{13}$$

$$u = 70$$

27. Solve:

$$\frac{x}{21} = 34$$

$$\frac{x}{21} \cdot 21 = 34 \cdot 21$$

$$x = 714$$

29. Solve:

$$19w = 665$$

$$\frac{19w}{19} = \frac{665}{19}$$

$$w = 35$$

31. Solve:

$$\frac{r}{56} = 51$$

$$\frac{r}{56} \cdot 56 = 51 \cdot 56$$

$$r = 2856$$

33. Solve:

$$31a = 775$$

$$\frac{31a}{31} = \frac{775}{31}$$

$$a = 25$$

35. Solve:

$$\frac{y}{48} = 9$$

$$\frac{y}{48} \cdot 48 = 9 \cdot 48$$

$$y = 432$$

37. Solve:

$$168 = 24x$$

$$\frac{168}{24} = \frac{24x}{24}$$

$$7 = x$$

39. Solve:

$$168 = \frac{x}{24}$$

$$168 \cdot 24 = \frac{x}{24} \cdot 24$$

$$4032 = x$$

41. If $4 \cdot w = 76$, then $w = 19$

43. The correct statement is: If $5x = 100$, then $x = 20$. To solve this equation correctly, divide both sides by 5, rather than subtracting 5 from both sides.

45. The wholesale cost:

$$\$5580 = 18p$$

$$\frac{\$5580}{18} = \frac{18p}{18}$$

$$\$310 = p$$

47. The number of cans sold:

$$N = 24c$$

$$N = 24(\$723)$$

$$N = \$17,352$$

49. The number of bricks:

$$N = 32d$$

$$N = 32(212)$$

$$N = 6784 \text{ bricks}$$

51. The perimeter of the square:

$$100\,\text{m} = 4s$$

$$\frac{100\,\text{m}}{4} = \frac{4s}{4}$$

$$25\,\text{m} = s$$

53. Average daily high Jan. temp is:

$y = $ July temp

$x = $ Jan temp

$$y = 2x$$

$$60° = 2x$$

$$\frac{60°}{2} = \frac{2x}{2}$$

$$30°F = x$$

55. The height of the screen is:

$$14\,\text{ft} = 2h$$

$$\frac{14\,\text{ft}}{2} = \frac{2h}{2}$$

$$7\,\text{ft} = h$$

57. The meeting planner objected because the screen is too short:

$$16\,\text{ft} = 2h$$

$$\frac{16\,\text{ft}}{2} = \frac{h}{2}$$

$$8\,\text{ft} = h$$

He recommended a 9 ft screen:

$$72\,\text{ft} = 8h$$

$$\frac{72\,\text{ft}}{8} = \frac{8h}{8}$$

$$9\,\text{ft} = h$$

59. The equations $\frac{x}{2} = 507$ and $x = 700$ are both equivalent to $x = 100$.

The equations $\frac{x}{2} = 50$ and $0\left(\frac{x}{2}\right) = 0(50)$ are not equivalent because the first

equation is equivalent to $x = 100$, whereas the second equation is equivalent to $0 = 0$.

61. Solve:

$$44x = 17{,}996$$
$$\frac{44x}{44} = \frac{17{,}996}{44}$$
$$x = 409$$

63. Solve:

$$\frac{y}{707} = 25$$
$$\frac{y}{707} \cdot 707 = 25 \cdot 707$$
$$y = 17{,}675$$

65. Solve:

$$\frac{7t}{5} = 406$$
$$\frac{7t}{5} \cdot 5 = 406 \cdot 5$$
$$7t = 2030$$
$$\frac{7t}{7} = \frac{2030}{7}$$
$$t = 290$$

67. Seven more than the product of two numbers: $7 + xy$

69. The quotient of the second number and 17 diminished by the first number: $\frac{y}{17} - x$

71. The product of two numbers less their quotient: $xy - \frac{x}{y}$

Section 3.7
Solving Equations of the Form $ax + b = c$ or $ax - b = c$

1. Solve:
$$4x + 1 = 9$$
$$4x + 1 - 1 = 9 - 1$$
$$4x = 8$$
$$\frac{4x}{4} = \frac{8}{4}$$
$$x = 2$$

3. Solve
$$6y - 3 = 27$$
$$6y - 3 + 3 = 27 + 3$$
$$6y = 30$$
$$\frac{6y}{6} = \frac{30}{6}$$
$$y = 5$$

5. Solve:
$$10w + 8 = 88$$
$$10w + 8 - 8 = 88 - 8$$
$$10w = 80$$
$$\frac{10w}{10} = \frac{80}{10}$$
$$w = 8$$

7. Solve:
$$9t + 5 = 5$$
$$9t + 5 - 5 = 5 - 5$$
$$9t = 0$$
$$\frac{9t}{9} = \frac{0}{9}$$
$$t = 0$$

9. Solve:
$$4z - 8 = 20$$
$$4z - 8 + 8 = 20 + 8$$
$$4z = 28$$
$$\frac{4z}{4} = \frac{28}{4}$$
$$z = 7$$

11. Solve:
$$8a - 7 = 57$$
$$8a - 7 + 7 = 57 + 7$$
$$8a = 64$$
$$\frac{8a}{8} = \frac{64}{8}$$
$$a = 8$$

13. Solve:
$$5n + 6 = 41$$
$$5n + 6 - 6 = 41 - 6$$
$$5n = 35$$
$$\frac{5n}{5} = \frac{35}{5}$$
$$n = 7$$

15. Solve:
$$34 = 9p - 11$$
$$34 + 11 = 9p - 11 + 11$$
$$45 = 9p$$
$$\frac{45}{9} = \frac{9p}{9}$$
$$5 = p$$

17. Solve:
$$6a - 24 = 90$$
$$6a - 24 + 24 = 90 + 24$$
$$6a = 114$$
$$\frac{6a}{6} = \frac{114}{6}$$
$$a = 19$$

19. Solve:
$$12c + 29 = 53$$
$$12c + 29 - 29 = 53 - 29$$
$$12c = 24$$
$$\frac{c}{12} = \frac{24}{12}$$
$$c = 2$$

21. Solve:
$$18 + 13x = 96$$
$$18 - 18 + 13x = 96 - 18$$
$$13x = 78$$
$$\frac{13x}{13} = \frac{78}{13}$$
$$x = 6$$

23. Solve:
$$12y - 56 = 28$$
$$12y - 56 + 56 = 28 + 56$$
$$12y = 84$$
$$\frac{12y}{12} = \frac{84}{12}$$
$$y = 7$$

25. Solve:
$$31t - 45 = 730$$
$$31t - 45 + 45 = 730 + 45$$
$$31t = 775$$
$$\frac{31t}{31} = \frac{775}{31}$$
$$t = 25$$

27. Solve:
$$407 = 24a + 215$$
$$407 - 215 = 24a + 215 - 215$$
$$192 = 24a$$
$$\frac{192}{24} = \frac{24a}{24}$$
$$8 = a$$

29. Solve:
$$42 = 35b - 98$$
$$42 + 98 = 35b - 98 + 98$$
$$140 = 35b$$
$$\frac{140}{35} = \frac{35b}{35}$$
$$4 = b$$

31. Solve:
$$974 = 232 + 53c$$
$$974 - 232 = 232 - 232 + 53c$$
$$742 = 53c$$
$$\frac{742}{53} = \frac{53c}{53}$$
$$14 = c$$

33. Solve:
$$2y + 3y - 65 = 210$$
$$5y - 65 = 210$$
$$5y - 65 + 65 = 210 + 65$$
$$5y = 275$$
$$\frac{5y}{5} = \frac{275}{5}$$
$$y = 55$$

35. Solve:
$$7w + 14 + 14w = 140$$
$$21w + 14 = 140$$
$$21w + 14 - 14 = 140 - 14$$
$$21w = 126$$
$$\frac{21w}{21} = \frac{126}{21}$$
$$w = 6$$

37. Solve:
$$234 + 50x + 7x = 1545$$
$$234 + 57x = 1545$$
$$234 - 234 + 57x = 1545 - 234$$
$$57x = 1311$$
$$\frac{57x}{57} = \frac{1311}{57}$$
$$x = 23$$

39. Solve:
$$1005 = 104z - 23z + 114$$
$$1005 = 81z + 114$$
$$1005 - 114 = 81z + 114 - 114$$
$$891 = 81z$$
$$\frac{891}{81} = \frac{81z}{81}$$
$$11 = z$$

41. If $a \cdot 6 + 9 = 21$, then $a = 2$
$$a \cdot 6 + 9 = 21$$
$$6a + 9 - 9 = 21 - 9$$
$$6a = 12$$
$$\frac{6a}{6} = \frac{12}{6}$$
$$a = 2$$

43. The correct statement is:
$$7x + 35 = 126$$
$$7x + 35 - 35 = 126 - 35$$
$$7x = 91$$
Subtract 35 from both sides,
rather than dividing both sides by 7.

45. The number is:
$$5n + 6 = 41$$
$$5n + 6 - 6 = 41 - 6$$
$$5n = 35$$
$$\frac{5n}{5} = \frac{35}{5}$$
$$n = 7$$

47. The number is:
$$92 = 4n - 12$$
$$92 + 12 = 4n - 12 + 12$$
$$104 = 4n$$
$$\frac{104}{4} = \frac{4n}{4}$$
$$26 = n$$

49. The number of units is:

$$n + 3n = 1024$$
$$4n = 1024$$
$$\frac{4n}{4} = \frac{1024}{4}$$
$$n = 256 \ \text{(manual line)}$$
$$3n = 768 \ \text{(automated line)}$$

51. The required height of the box is:
$$s = 2wl + 2wh + 2hl$$
$$1510 = 2(15)(20) + 2(15)h + 2(20)h$$
$$1510 = 600 + 30h + 40h$$
$$1510 = 600 + 70h$$
$$1510 - 600 = 600 - 600 + 70h$$
$$910 = 70h$$
$$\frac{910}{70} = \frac{70h}{70}$$
$$13 \ \text{in} = h$$

53. ATT: $3000 = 0 + 15m, \ m = 200$
Tone: $3000 = 490 + 10m, \ m = 250$
Pace: $3000 = 696 + 9m, \ m = 292.3$

Pace is the company that will give Jessica the most minutes for her \$30.

55. The width of the room is:
$$72 + 18x + 72 + 18x + 72 = 1200$$
$$36x + 216 = 1200$$

57. answers vary

59. Solve:
$$4w - 54 = 22$$
$$4w - 54 + 54 = 22 + 54$$
$$4w = 76$$
$$\frac{4w}{4} = \frac{76}{4}$$
$$w = 19$$

61. Solve:
$$19x - 554 = 10{,}979$$
$$19x - 554 + 554 = 10{,}979 + 554$$
$$19x = 11{,}533$$
$$\frac{19x}{19} = \frac{11{,}533}{19}$$
$$x = 607$$

63. Solve:

$$\frac{z}{9} - 626 = 7$$

$$\frac{z}{9} - 626 + 626 = 7 + 626$$

$$\frac{z}{9} = 633$$

$$\frac{z}{9} \cdot 9 = 633 \cdot 9$$

$$z = 5697$$

65. Solve:

$$ax + b = c$$

$$ax + b - b = c - b$$

$$ax = c - b$$

$$\frac{ax}{a} = \frac{c - b}{a}$$

$$x = \frac{c - b}{a}$$

67.

$$\frac{\$523 + \$397 + \$468 + \$504 + x}{5} = \frac{\$2500}{5} = \$500$$

$$\frac{\$1892 + x}{5} = \$500$$

$$\$1892 + x = \$2500$$

$$x = \$608$$

69. Evaluate:

$$800 - 8^2 \left(2^3\right)$$

$$800 - 64(8)$$

$$800 - 512$$

$$288$$

71. Perimeter: $12 \text{ ft} + 3 \text{ ft} + 8 \text{ ft} + 13 \text{ ft} = 36 \text{ ft}$

73. Volume: $66 \text{ ft}^2 \left(8 \text{ ft}\right) = 528 \text{ ft}^3$

Chapter 3
True-False Concept Review

1. False – A variable is a changing letter.

2. True

3. True

4. True

5. True

6. True

7. True

8. True

9. False – The algebraic expressions $5ab$ and $5xy$ are unlike terms.

10. True

11. True

12. False – Unlike terms cannot be combined. Like terms can be combined.

13. True

14. False – The exponents of terms are not changed whenever like terms are added.

15. False – When like terms are subtracted, the exponents of the variables remain the same.

16. False – Equivalent equations have the same solutions.

17. True

18. False – To check an equation, we must substitute a value for the variable in the equation.

19. True

20. False – An algebraic expression does not contain an equal sign.

21. False – The sum of 6 and 4 is 10.

22. False – The square of 6 is 36.

23. True

24. False – For all values of x and y, $6x + 3y$ cannot be added.

25. True

Chapter 3
Review

1. A number increased by 30: $x + 30$

3. Fifty-six more than four times a number: $56 + 4x$

5. Ninety-five divided by the product of seven and a number: $\dfrac{95}{7x}$

7. $45 - b = 45 - 23 = 22$

9. $x^2 - 16 = 8^2 - 16 = 64 - 16 = 48$

11.   4 is a solution
$$19y - 29 = 47$$
$$19(4) - 29 = 47$$
$$76 - 29 = 47$$
$$47 = 47$$

13.   19 is a solution
$$6x - 45 = 69$$
$$6(19) - 45 = 69$$
$$114 - 45 = 69$$
$$69 = 69$$

15. 23 is not a solution
$$17y + 24 = 414$$
$$17(23) + 24 = 414$$
$$391 + 24 = 414$$
$$415 \neq 414$$

17. Evaluate:
$$V = s^3$$
$$V = 7^3$$
$$V = 343$$

19. Evaluate:
$$S = c + m$$
$$S = 510 + 23$$
$$S = 533$$

21. Solve:
$$a - 13 = 41$$
$$a - 13 + 13 = 41 + 13$$
$$a = 54$$

23. Solve:
$$2x - 6 = 22$$
$$2x - 6 + 6 = 22 + 6$$
$$2x = 28$$
$$\frac{2x}{2} = \frac{28}{2}$$
$$x = 14$$

25. Solve:
$$2y + 12 + 3y = 67$$
$$5y + 12 = 67$$
$$5y + 12 - 12 = 67 - 12$$
$$5y = 55$$
$$\frac{5y}{5} = \frac{55}{5}$$
$$y = 11$$

27. $43t - 21t + 9t = 31t$

29. $4x^2 + 3x + 19x^2 - x = 23x^2 + 2x$

31. $(3x + 9y) + (4x - 5y) = 7x + 4y$

33. $(7x + 12y + 4z) + (2x - 3y) = 9x + 9y + 4z$

35. $(2a + 3b + c) - (a + b + c) = 2a + 3b + c - a - b - c = a + 2b$

37. $(24z)(9y) = 216zy$

39. $\dfrac{56ab}{8} = 7ab$

41. $3a(4a - 7) = 12a^2 - 21a$

43. $(2x + 5)(x + 4) = 2x^2 + 13x + 20$

45. $\dfrac{153x^2y + 204xy^2 - 289xy}{17xy} = 9x + 12y - 17$

47. Solve:

$$t - 45 = 80$$
$$t - 45 + 45 = 80 + 45$$
$$t = 125$$

49. Solve:

$$s + 7 = 75$$
$$s + 7 - 7 = 75 - 7$$
$$s = 68$$

51. Solve:

$$9t = 63$$
$$\frac{9t}{9} = \frac{63}{9}$$
$$t = 7$$

53. Solve:

$$195 = 13y$$
$$\frac{195}{13} = \frac{13y}{13}$$
$$15 = y$$

55. Solve:

$$\frac{x}{7} = 14$$
$$\frac{x}{7} \cdot 7 = 14 \cdot 7$$
$$x = 98$$

57. Solve:

$$\frac{w}{12} = 17$$
$$\frac{w}{12} \cdot 12 = 17 \cdot 12$$
$$w = 204$$

59. The cost of one umbrella is:

$$62x = \$434$$
$$\frac{62x}{62} = \frac{\$434}{62}$$
$$x = \$7$$

61. Solve:

$$41 + 5x = 76$$
$$41 - 41 + 5x = 76 - 41$$
$$5x = 35$$
$$\frac{5x}{5} = \frac{35}{5}$$
$$x = 7$$

63. Solve:

$$22w + 14 = 146$$
$$22w + 14 - 14 = 146 - 14$$
$$22w = 132$$
$$\frac{22w}{22} = \frac{132}{22}$$
$$w = 6$$

65. The number is:

$$27 + 6n = 129$$
$$27 - 27 + 6n = 129 - 27$$
$$6n = 102$$
$$\frac{6n}{6} = \frac{102}{6}$$
$$n = 17$$

Chapter 3
Test

1. $32z - 19z + 48z = 61z$

2. Solve:
    $$5x + 28 = 108$$
    $$5x + 28 - 28 = 108 - 28$$
    $$5x = 80$$
    $$\frac{5x}{5} = \frac{80}{5}$$
    $$x = 16$$

3. Evaluate:
    $$336 - 18y$$
    $$336 - 18(13)$$
    $$336 - 234$$
    $$102$$

4. $S = \dfrac{n(a+l)}{2} = \dfrac{20(2+59)}{2} = 10(61) = 610$

5. Solve:
    $$w - 336 = 792$$
    $$w - 336 + 336 = 792 + 336$$
    $$w = 1128$$

6. Solve:
    $$\frac{t}{21} = 6$$
    $$\frac{t}{21} \cdot 21 = 6 \cdot 21$$
    $$t = 126$$

7. $\dfrac{180ab}{15b} = 12a$

8. $\dfrac{100c^2 + 60c - 40cd}{20c} = 5c + 3 - 2d$

9. $3xy(14x + 23y - 15z) = 42x^2y + 69xy^2 - 45xyz$

10. $215x - 17x + 29x = 227x$

11. Evaluate:
    $$3p + pq - q$$
    $$3(45) + 45(37) - 37$$
    $$135 + 1665 - 37$$
    $$1763$$

12. Solve:
    $$r + 79 = 107$$
    $$r + 79 - 79 = 107 - 79$$
    $$r = 28$$

13. $(16x)(4x)(x) = 64x^3$

14. Solve:

$$24y = 264$$

$$\frac{24y}{24} = \frac{264}{24}$$

$$y = 11$$

15. Solve:

$$4x + 19 = 91$$

$$4x + 19 - 19 = 91 - 19$$

$$4x = 72$$

$$\frac{4x}{4} = \frac{72}{4}$$

$$x = 18$$

16. $25y - y - 7y = 17y$

17. Solve:

$$12x - 168 = 168$$

$$12x - 168 + 168 = 168 + 168$$

$$12x = 336$$

$$\frac{12x}{12} = \frac{336}{12}$$

$$x = 28$$

18. 0 is not a solution

$$191x - 382 = 382$$

$$191(0) - 382 = 382$$

$$0 - 382 = 382$$

$$-382 \neq 382$$

19. $\dfrac{18 + n}{5}$

20. $C = np = 35(16) = 560$

21. The bedroom set costs:

$$C = d + pm$$

$$\$76 + \$21(24)$$

$$\$76 + \$504$$

$$\$580$$

22. The length is:

$$204\,\text{ft} = 2l + 2(19\,\text{ft})$$

$$204\,\text{ft} = 2l + 38\,\text{ft}$$

$$166\,\text{ft} = 2l$$

$$83\,\text{ft} = l$$

23. The cost of one lawnmower is:

$$18c = \$6264$$

$$\frac{18c}{18} = \frac{\$6264}{18}$$

$$c = \$348$$

24. The number is:

$$374 - 14n = 10$$

$$374 - 10 - 14n = 10 - 10$$

$$364 - 14n = 0$$

$$164 - 14n + 14n = 0 + 14n$$

$$164 = 14n$$

$$\frac{164}{14} = \frac{14n}{14}$$

$$26 = n$$

CUMULATIVE REVIEW
CHAPTERS 1-3

1. eight thousand, five = 8,005

3. 168 > 129

5. 987 + 677 + 9656 + 28 = 11,348

7. 8235 − 2348 = 5887 rounded to the nearest ten is 5890

9. 3456(209) = 722,304 rounded to the nearest thousand is 722,000

11. 101,595 ÷ 195 = 521

13. $12^3 = 1728$

15. $28,040,000 ÷ 10^3 = 28,040$

17. $\left(9^3\right)^4 = 9^{3\cdot4} = 9^{12}$

19. Simplify:
$$11^2 + 8^2 ÷ 2^2 - \left(3^2\right)^2$$
$$121 + 64 ÷ 4 - 81$$
$$121 + 16 - 81$$
$$56$$

21. $\left(11+8\right)^2 = 19^2 = 361$

23. $\dfrac{234 + 486 + 126 + 596 + 222 + 196}{6} = \dfrac{1860}{6} = 310$

25. The apartments were never all completely occupied.

27. The decline in occupancy between 1995 and 1996 is: 475 − 400 = 75
    If the decline is also 75 in 1997, then the 1997 occupancy will be 400 − 75 = 325

29. The line graph is:

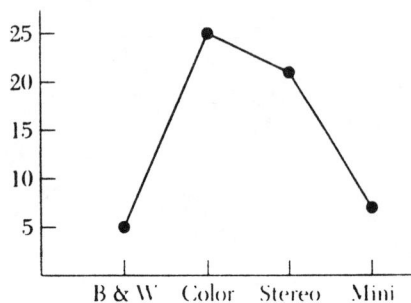

31. Subtract:

       3 hr 25 min 41 sec

     − 1 hr 30 min 54 sec

33. $\dfrac{12\,\text{lb} + 22\,\text{lb} + 35\,\text{lb} + 28\,\text{lb} + 45\,\text{lb} + 62\,\text{lb}}{6} = \dfrac{204\,\text{lb}}{6} = 34\,\text{lb}$

35. The perimeter is:

$2(18\,\text{ft}) + 2(12\,\text{ft}) + 2(4\,\text{ft}) + 2(9\,\text{ft}) = 36\,\text{ft} + 24\,\text{ft} + 8\,\text{ft} + 18\,\text{ft} = 86\,\text{ft}$

37. $(16\,\text{ft})(21\,\text{ft}) \div 125\,\text{ft}^2 = 336\,\text{ft}^2 \div 125\,\text{ft}^2 / \text{gal} = 2.6\,\text{gal} \approx 3\,\text{gal}$

39. The product of 11 and a number and that product decreased by 9: $11n - 9$

41. Evaluate:

$x^2 + 4x + 2$

$6^2 + 4(6) + 2$

$36 + 24 + 2$

62

43. 9 is a solution.

$4x - 5 = 31$

$4(9) - 5 = 31$

$36 - 5 = 31$

$31 = 31$

45. Evaluate:

$P = 2l + 2w$

$P = 2(12) + 2(9)$

$P = 24 + 18$

$P = 42$

47. $8x + 9x + 17x = 34x$

49. $a + 9a - 6a = 4a$

51. $21x + 17y + 9x = 30x + 17y$

53. $2x(3x + 2y - 3z) = 6x^2 + 4xy - 6xz$

55. $12a(9a)(4a) = 432a^3$

57. $\dfrac{136a^2bc}{17ab} = 8ac$

59. Solve:

$942 = m - 78$

$942 + 78 = m - 78 + 78$

$1020 = m$

61. Solve:

$$\frac{t}{28} = 35$$

$$\frac{t}{28} \cdot 28 = 35 \cdot 28$$

$$t = 980$$

63. Solve:

$$5x - 9 = 46$$

$$5x - 9 + 9 = 46 + 9$$

$$5x = 55$$

$$\frac{5x}{5} = \frac{55}{5}$$

$$x = 11$$

65. Solve:

$$12a + 15 = 39$$

$$12a + 15 - 15 = 39 - 15$$

$$12a = 24$$

$$\frac{12a}{12} = \frac{24}{12}$$

$$a = 2$$

67. The area of the landscaping is:

$$60\,\text{ft}(150\,\text{ft}) + 150\,\text{ft}(210\,\text{ft}) - 185\,\text{ft}(120\,\text{ft})$$

$$9000\,\text{ft}^2 + 31{,}500\,\text{ft}^2 - 22{,}200\,\text{ft}^2$$

$$40{,}500\,\text{ft}^2 - 22{,}200\,\text{ft}^2$$

$$18{,}300\,\text{ft}^2$$

CHAPTER FOUR
INTEGERS and EQUATIONS

Section 4.1
Opposites, Absolute Value, and Inequalities

1. $-(-3) = +3$

3. $-(14) = -14$

5. $-(-37) = +37$

7. $-(17) = -17$

9. The opposite of -61 is 61.

11. The opposites of -75, 81 and 95 are 75, -81 and -95.

13. The opposites of -65, -82 and -67 are 65, 82 and 67.

15. $-[-(-34)] = -34$

17. $-[-(-75)] = -75$

19. $|-11| = +11$

21. $|33| = 33$

23. $|-172| = +172$

25. $|144| = 144$

27. The absolute value of both ± 32 is 32.

29. The absolute values of -75, 81 and 100 are 75, 81 and 100.

31. The absolute values of 99, -45 and -115 are 99, 45 and 115.

33. $|-23| - |17| = 23 - 17 = 6$

35. $|88 - 45| = |43| = 43$

37. $-6 < 3$

39. $-19 > -28$

41. $-35 > -76$

43. $19 > -67$

45. -13 is greater than -55

47. $-(-11) < 12$

49. $-(16-5) < -(9)$

51. $-(-23) < -(-45)$

53. $-(12-9) > -(11-6)$

55. $-(-(-(-34))) = 34$

57. $-[|-75|-|-45|] = -30$

59. $23-12 < 34-22$ is true

61. The opposite of a gain of (+$2) is a loss of (−$2).

63. The opposite of a reading of 12°C is $-12°C$.

65. The opposite of 1875 BC or −1875 is 1875 AD or +1875.

67. Solve:
$$|x| = 11$$
$$x = \pm 11$$

69. Solve:
$$|x| = 101$$
$$x = \pm 101$$

71. The lowest temperature recorded during the five days was $-115°\,C$.

73. The lowest temperature recorded at 6:00 PM is $-102°\,C$.

75. The continental altitudes in the application written as integers are:

Continent	Highest Point	Lowest Point
Africa	+19,340	−512
Antarctica	+16,864	−8327
Asia	+29,028	−1312
Australia	+7,310	−52
Europe	+18,510	−92
North America	+20,320	−282
South America	+22,834	−131

77. The highest point on the Earth (29,028 ft) is closer to sea level than the deepest Point in the ocean (35,840 ft).

79. The absolute value of each number in the set of "positive numbers and zero" is the number itself.

81. answers vary

83. Simplify:
$$8 - |12 - 8| - |10 - 8| + 2$$
$$8 - |4| - |2| + 2$$
$$8 - 4 - 2 + 2$$
$$4$$

85. If n is a negative number, then $-n$ is a positive number.

87. answers vary

89. $37^7 \cdot 37^2 \cdot 37^{11} = 37^{7+2+11} = 37^{20}$

91. The average is: $\dfrac{234 + 176 + 89 + 421 + 1055}{5} = \dfrac{1975}{5} = 395$

Section 4.2
Adding Integers

1. $-8 + 2 = -6$

3. $6 + (-7) = -1$

5. $-8 + (-6) = -14$

7. $(-17) + (-17) = -34$

9. $7 + (-15) = -8$

11. $23 + (-18) = 5$

13. $(-11) + (-23) = -34$

15. $5 + (-14) = -9$

17. $48 + (-39) = 9$

19. $-38 + (-57) = -95$

21. $(-56) + 56 = 0$

23. $-3 + (-6) + 5 = -4$

25. $-18 + 21 + (-3) = 0$

27. $47 + (-32) + (-14) = 1$

29. $(-94) + 61 = -33$

31. $310 + (-420) = -110$

33. $c = 45 + (-82) = -37$

35. Solve:
$$a - 11 = -9$$
$$a - 11 + 11 = -9 + 11$$
$$a = 2$$

37. $135 + (-256) = -121$

39. $-81 + (-32) + (-76) = -189$

41. $-31 + 28 + (-63) + 36 = -94 + 64 = -30$

43. $49 + (-67) + 27 + 72 = 148 - 67 = 81$

45. $356 + (-762) + (-892) + 541 = 897 + (-1654) = -757$

47. $542 + (-481) + (-175) = 542 + (-656) = -114$

49. $(-57°) + (-49°) + (-86°) + (-93°) + (-64°) = -349°C$

51. $x + (-67)$

53. $[22 + (-12)] + (-9) = 10 - 9 = 1$

55. $-877 \, lb + 764 \, lb = -113 \, lb$

57. $(+30{,}000) + (+220) + (-200) + (+55) + (-110) + (-55) + (-84) = +29{,}826$

59. $34{,}873 + 2386 - 11{,}705 - 875 - 563 + 5{,}075 - 723 = 28{,}468$ volumes

61. $-8 \, yd + 10 \, yd - 3 \, yd + 9 \, yd = +19 \, yd - 11 \, yd = +8 \, yd$
Since +10 yd are needed for a first down, +8 yd is not enough.

63. $-\$56 + \$18 - \$87 + \$112 = -\$143 + \$130 = -\$13$ (loss)

65. When adding a positive and negative number, take the difference of the absolute values of the two numbers and assign the sign of the number with the larger absolute value to the sum.

67. $(+15) + (-32) = -17$

69. Simplify:
$$|73 + (-59)| + |-43 + (-99) + (-12)| + +(-81)$$
$$|14| + |-154| + (-81)$$
$$14 + 154 - 81$$
$$168 - 81$$
$$87$$

71. Solve:
$$|x + 6 + (-11)| = 5 + (-45)$$
$$|x - 5| = -40$$

There is no solution to this equation, since it is impossible for an absolute value to be negative.

73. $9789 \times 10^8 = 978{,}900{,}000{,}000$

75. $(26^{14})(26^9) = 26^{14+9} = 26^{23}$

77. $\dfrac{65}{4x}$

Section 4.3
Subtracting Integers

1. $-6 - 4 = -6 + (-4) = -10$

3. $-5 - (-8) = -5 + 8 = 3$

5. $-12 - 5 = -12 + (-5) = -17$

7. $17 - (-15) = 17 + 15 = 32$

9. $-14 - 12 = -14 + (-12) = -26$

11. $-16 - (-17) = -16 + 17 = 1$

13. $-16 - (-16) = -16 + 16 = 0$

15. $-32 - 22 = -32 + (-22) = -54$

17. $56 - (-31) = 56 + 31 = 87$

19. $-65 - 73 = -65 + (-73) = -138$

21. $-65-(-69) = -65 + 69 = 4$

23. $-74-(-74) = -74 + 74 = 0$

25. $145-(-32) = 145 + 32 = 177$

27. $43-(-73) = 43 + 73 = 116$

29. $-349-328 = -349 + (-328) = -677$

31. -12 is a solution
$$x - 12 = -24$$
$$-12 - 12 = -24$$
$$-24 = -24$$

33. Simplify:
$$-7 - (-9) - 12$$
$$-7 + (+9) - 12$$
$$-10$$

35. Simplify:
$$-18 - (-16) - (-7)$$
$$-18 + (+16) + (+7)$$
$$-18 + 23$$
$$5$$

37. Solve:
$$b + 56 = -34$$
$$b + 56 - 56 = -34 - 56$$
$$b = -90$$

39. Solve:
$$d + 19 = 7$$
$$d + 19 - 19 = 7 - 19$$
$$d = -12$$

41. Simplify:
$$-43 - 23 - 21$$
$$-43 + (-23) + (-21)$$
$$-87$$

43. Simplify:
$$56 - 29 - 65$$
$$56 + (-29) + (-65)$$
$$56 + (-94)$$
$$-38$$

45. Simplify:
$$-47 - 71 - (-61)$$
$$-47 + (-71) + (+61)$$
$$-118 + (+61)$$
$$-57$$

47. Simplify:
$$-43 - (-24) - 11 - (-28)$$
$$-43 + (+24) + (-11) + (+28)$$
$$-54 + 52$$
$$-2$$

49. $\$25,250 - \$27,950 = -\$2700$

51. $-38° - (-135°) = -38° + (+135°) = 97°\,C$

53. $(12 - 23) - 32 = -11 - 32 = -11 + (-32) = -43$

55. The range of altitude for each continent is:

Africa:	$19,340 + 512 = 19,852$
Antarctica:	$16,864 + 8327 = 25,191$
Asia:	$29,028 + 1312 = 30,340$
Australia:	$7,310 + 52 = 7,362$
Europe:	$18,510 + 92 = 18,602$
North America:	$20,320 + 282 = 20,602$
South America:	$22,834 + 131 = 22,965$

The continent with the smallest range is Australia, which means that the terrain of the continent is relatively flat.

57. The depth of the ocean floor is: $-33,476\,\text{ft} + 13,796\,\text{ft} = -19,680\,\text{ft}$

59. Thomas's account balance is: $\$117 - \$188 = -\$71$

61. $-105° - (-90°) = -15°\,\text{C}$

63. $-10° - (-115°) = 105°\,\text{C}$

65. answers vary

67. answers vary

69. Simplify:

$$-34 - (-42) - |-(-32) - (-21)|$$
$$-34 + 42 - |32 + 21|$$
$$8 - 53$$
$$-45$$

71. Solve:
$$|a - (-11)| = 14$$

$$a + 11 = 14 \text{ or } a + 11 = -14$$
$$a = 3 \text{ or } a = -25$$

73. answers vary

75. Simplify:

$$(28 + 3 \cdot 4) - 2 \cdot 3 + 4$$
$$(28 + 12) - 6 + 4$$
$$40 - 6 + 4$$
$$34 + 4$$
$$38$$

77. Evaluate:
$$b^3 - 4b^2 - 10$$
$$5^3 - 4(5)^2 - 10$$
$$125 - 4(25) - 10$$
$$125 - 100 - 10$$
$$25 - 10$$
$$15$$

Section 4.4
Adding and Subtracting Polynomials (Integers)

1. $5x + 9x = 14x$

3. $-11x + 8x = -3x$

5. $8a + (-6a) = 2a$

7. $-17a - 13a = -17a + (-13a) = -30a$

9. $-13y + (-5y) = -18y$

11. $-12y - (-9y) = -12y + (+9y) = -3y$

13. $4a + 6a + (-5a) = 10a + (-5a) = 5a$

15. $7x + (-6x) + (-9x) = 7x + (-15x) = -8x$

17. $3x^2 - 8x^2 + 7x^2 = 10x^2 - 8x^2 = 2x^2$

19. $(3a - 7) + (-4a + 8) = -a + 1$

21. $(x^2 - 8x) + (-5x^2 + 6x) = -4x^2 - 2x$

23. $(4x - 17) - (7x + 9) = 4x - 17 - 7x - 9 = -3x - 26$

25. $9xy + 12xy - 8xy + (-24xy) - (-15xy) = 9xy + 12xy - 8xy - 24xy + 15xy = 4xy$

27. $(8y^2 - 4y - 7) + (3y^2 + 7y - 5) = 11y^2 + 3y - 12$

29.
$$\begin{array}{r} 3a - 2b + 9 \\ +\ -7a - 3b + 11 \\ \hline -4a - 5b + 20 \end{array}$$

31.
$$\begin{array}{r} 7c - 8d - 3 \\ -(-2c - 5d - 6) \\ \hline 9c - 3d + 3 \end{array}$$

33. $3a + [5a + (-7a)] = 3a + (-2a) = a$

35. $[(-11b) + (-7b)] - 15b = (-18b) + (-15b) = -33b$

37. $(5x^2 - 9x - 11) - (3x^2 + 8x - 17) = 5x^2 - 9x - 11 - 3x^2 - 8x + 17 = 2x^2 - 17x + 6$

39. Simplify:
$$(3a - 2b + 7) + (2a - 8b - 9) - (4a + 6b + 3)$$
$$3a - 2b + 7 + 2a - 8b - 9 - 4a - 6b - 3$$
$$a - 16b - 5$$

41. $(2a + 5b) - (4a - 3b) = 2a + 5b - 4a + 3b = -2a + 8b$

43. .Combine:
$$(3x + 7y - 9) + (6x + 3y - 8) - (4x - 7y - 2)$$
$$3x + 7y - 9 + 6x + 3y - 8 - 4x + 7y + 2$$
$$5x + 17y - 15$$

45. Ben: $3c + 4t$ Josh: $5c + t$ $(5c + t) - (3c + 4t) = 5c + t - 3c - 4t = 2c - 3t$
No decision can be made since the actual costs are not known.

47. $(3x^2 - 5x + 6) - (10x - 4) = 3x^2 - 15x + 10$

49. answers vary

51. answers vary

53. Simplify:
$$3(2a - 7b) + 2(3a + 2b + 6) + 7(4a - 7)$$
$$6a - 21b + 6a + 4b + 12 + 28a - 49$$
$$40a - 17b - 37$$

55. answers vary

57. 11 is a solution
$$5y - 32 = 23$$
$$5(11) - 32 = 23$$
$$55 - 32 = 23$$
$$23 = 23$$

59. $8(3x - 12) = 24x - 96$

Section 4.5
Multiplying and Dividing Integers

1. The product of two negative integers is <u>positive</u>.

3. $(-3)(2) = -6$

5. $(-5)(-6) = 30$

7. $6(-8) = -48$

9. $(-12)(-3) = 36$

11. $(-5)(-11) = 55$

13. $(-11)(10) = -110$

15. $(-12)(-12) = 144$

17. $(43)(-6) = -258$

19. $(-14)(-15) = 210$

21. $(-7)(-2)(-3) = -42$

23. $(-8)(-2)(3) = 48$

25. $(-1)(-1)(-1)(-1)(-1)(-6) = 6$

27. $8 \div (-2) = -4$

29. $(-15) \div 3 = -5$

31. $\dfrac{-14}{-2} = 7$

33. $\dfrac{-33}{11} = -3$

35. $\dfrac{-66}{-11} = 6$

37. $\dfrac{-32}{-8} = 4$

39. $125 \div (-25) = -5$

41. $(-260) \div 13 = -20$

43. $\dfrac{-300}{-12} = 25$

45. $\dfrac{252}{-14} = -18$

47. $\dfrac{-918}{-9} = 102$

49. $\dfrac{-475}{25} = -19$

51. $\dfrac{594}{-18} = -33$

53. $39(-18) = -702$

55. $\dfrac{-384}{-24} = 16$

57. $(13)(-2)(-10)(5) = 1300$

59. $(-13)(-12)(-8)(2) = -2496$

61. $\dfrac{-560}{-28} = 20$

63. $\dfrac{-1071}{17} = -63$

65. $(-34)(-123)(65) = 271{,}830$

67. $\dfrac{-16{,}272}{36} = -452$

69. -15 is a solution

$$8x = -120$$
$$8(-15) = -120$$
$$-120 = -120$$

71. -18 is not a solution

$$\frac{z}{9} = -162$$
$$\frac{-18}{9} = -162$$
$$-2 \neq -162$$

73. $6(-4\,\text{lb}) = -24\,\text{lb}$

75. $\dfrac{-\$522}{6} = -\$87$

77. Asia's low point of -1312 ft. is about ten times the South America's low point of -131 ft.

79. Australia's high point of 7310 ft is about one-fourth Mt. Everest's height of 29,028 ft.

81. $7(-\$32) = -\$224$

83. $950(\$11) - 950(\$13) = \$10{,}450 - \$12{,}350 = -\$1900$

85. $\dfrac{-\$862{,}200}{20} = -\$43{,}110$ is the average monthly loss

$\dfrac{-\$862{,}200}{30} = -\$28{,}740$ is the total loss per stockholder

87. $-3^2 = -(3 \cdot 3) = -9$; whereas, $(-3)^2 = (-3)(-3) = 9$

89. $\dfrac{36 + 4}{-9} \neq -4 + 4$ because both the 36 and the 4 in the numerator of the fraction must be divided by the -9.

91. Simplify:
$$|-(-5)(-9-[-(-5)])|$$
$$|5|(-9-(+5))$$
$$5(-9-5)$$
$$5(-14)$$
$$-70$$

93. Simplify:
$$[-|-9|(8-12)] \div [(9-13)(8-7)]$$
$$[-(+9)(-4)] \div [(-4)(1)]$$
$$36 \div (-4)$$
$$-9$$

95. Evaluate:
$$6ab - a^2$$
$$6(-3)(-5) - (-3)^2$$
$$90 - 9$$
$$81$$

97. answers vary

99. $23s + 42t - 52s - 23t + 18s - 16t - 13s = -24s + 3t$

101. Solve:
$$y + 37 = 59$$
$$y + 37 - 37 = 59 - 37$$
$$y = 22$$

Section 4.6
Multiplying and Dividing Polynomials (Integers)

1. $7(-5c) = -35c$

3. $-8(-5d) = 40d$

5. $-7(-ab) = 7ab$

7. Multiply:

	$2x$	-5
-3	$-6x$	$+15$

9. $6(3t - 78) = 18t - 42$

11. $-2(3s - 7) = -6s + 14$

13. $3x(-2x) = -6x^2$

15. $-9b(-6b) = 54b^2$

17. $-11m(-12n) = 132mn$

19. $(-10b)(-8c) = 80bc$

21. $-4(4x - 5) = -16x + 20$

23. $4(-7x - 2) = -28x - 8$

25. $-13x(4x - 7) = -52x^2 + 91x$

27. $(x - 2)(x - 7) = x^2 - 9x + 14$

	x	-7
x	x^2	$-7x$
-2	$-2x$	$+14$

29. Multiply:
$(x + 6)(x - 5)$
$x^2 + 6x - 5x - 30$
$\quad x^2 + x - 30$

31. $12x \div (-4) = -3x$

33. $\dfrac{-25y}{-5} = 5y$

35. $\dfrac{-64b}{16} = -4b$

37. $\dfrac{-77ab}{7a} = -11b$

39. $105mn \div (-35m) = -3n$

41. The quotient is:

	$4x$	-3
3	$12x$	-9

43. $\dfrac{-120a^2}{-15a} = 8a$

45. $\dfrac{160m^2}{-20m} = -8m$

47. $\dfrac{2ab - 3ac + 5ad}{a} = 2b - 3c + 5d$

49. $\dfrac{3x^2 - 9x}{3x} = x - 3$

51. $(-18abc - 24bcd + 42bce) \div (-6bc) = 3a + 4d - 7e$

53. $\left(-123b^2 - 82b\right) \div -41b = 3b + 2$

55. $(b+9)(b+10) = b^2 + 10b + 9b + 90 = b^2 + 19b + 90$

57. $\dfrac{9x^2 - 15x}{3x} = 3x - 5$

59. $(y-7)(y+6) = y^2 + 6y - 7y - 42 = y^2 - y - 42$

61. $(c-8)(c+9) = c^2 + 9c - 8c - 72 = c^2 + c - 72$

63. $(3x - 13)(5x + 8) = 15x^2 + 24x - 65x - 104 = 15x^2 - 41x - 104$

65. $(3a - 2b)(7c - 8) = 21ac - 24a - 14bc + 16b$

67. $(5c - 1)(2c + 8) = 10c^2 + 40c - 2c - 8 = 10c^2 + 38c - 8$

69. $(3x + 9)(2x - 1) = 6x^2 - 3x + 18x - 9 = 6x^2 + 15x - 9$

71. Twentieth Century Securities

	Aggressive	Moderate	Conservative
Money Market	x	x	$3x$
Bonds	$4x$	$13x$	$9x$
Stocks	$15x$	$6x$	$8x$
Total	$20x$	$20x$	$20x$

a. The part of the aggressive fund in stocks is: $\dfrac{15x}{20x} = \dfrac{3}{4}$

b. The part of the moderate fund in money markets is: $\dfrac{x}{20x} = \dfrac{1}{20}$

c. The part of the conservative fund in bonds is: $\dfrac{9x}{20x} = \dfrac{9}{20}$

73. The area is:
$(15x + 4)(3x - 5) - 2(x + 2)$
$45x^2 - 75x + 12x - 20 - 2x - 4$
$45x^2 - 65x - 24$

75. Simplify:
$(3m - 7)(2m - 5)$
$6m^2 - 15m - 14m + 35$
$6m^2 - 29m + 35$

77. Simplify:

$$(x-9)(x+5)+(x+8)(x-2)$$

$$x^2+5x-9x-45+x^2-2x+8x-16$$

$$2x^2+2x-61$$

79. Simplify:

$$(2x+3)(2x-3)(x+6)$$

$$(4x^2-9)(x+6)$$

$$4x^3+24x^2-9x-54$$

81. The value of P is:

$$P=2L+2W$$

$$P=2(21)+2(13)$$

$$P=42+26$$

$$P=68$$

83. $13w(4wz)=52w^2z$

85. Solve:

$$78=m-37$$

$$78+37=m-37+37$$

$$115=m$$

Section 4.7
Order of Operations and Average (Integers)

1. Simplify:

$$-5(6)+(-22)$$

$$-30+(-22)$$

$$-52$$

3. Simplify:

$$34-7(-3)$$

$$34+21$$

$$55$$

5. Simplify:

$$-4\cdot6+7(-3)$$

$$-24+(-21)$$

$$-45$$

7. Simplify:

$$6-3^2$$

$$6-9$$

$$-3$$

9. Simplify:
$$-28 - (-3)(-4)$$
$$-28 - (+12)$$
$$-28 + (-12)$$
$$-40$$

11. The "squaring" operation is performed first.

13. Simplify:
$$5[10 + 3(-8)] - 33$$
$$5[10 - 24] - 33$$
$$5[-14] - 33$$
$$-70 - 33$$
$$-103$$

15. Simplify:
$$6(-10 + 4) - 33 \div (-11)$$
$$6(-6) - 33 \div (-11)$$
$$-36 + 3$$
$$-33$$

17. Simplify:

$$-35 \div (-5) + (-2)(-3)$$
$$7 + 6$$
$$13$$

19. Simplify:
$$8(-3) + 5^2 - 7(-3)$$
$$-24 + 25 + 21$$
$$22$$

21. Simplify:
$$(-5)^2 + (-2)^2 - (-5)(-2)$$
$$25 + 4 - 10$$
$$19$$

23. $\dfrac{18 - 18(2-3)}{-2(3)} = \dfrac{18 - 18(-1)}{-6} = \dfrac{18 + 18}{-6} = \dfrac{36}{-6} = -6$

25. $\dfrac{-5(6) - 3(8)}{6(-3) + 6(6)} = \dfrac{-30 - 24}{-18 + 36} = \dfrac{-54}{18} = -3$

27. $\dfrac{(-3) + (-4) + 6 + 5 + 1}{5} = \dfrac{5}{5} = 1$

29. $\dfrac{(-4) + (-7) + (-2) + (-3)}{4} = \dfrac{-16}{4} = -4$

31. $\dfrac{(-1) + (-2) + (-2) + (-3) + 5 + (-3)}{6} = \dfrac{-6}{6} = -1$

33. $\dfrac{(-8)+(-11)+(-14)+(-7)}{4} = \dfrac{-40}{4} = -10$

35. The missing number is:

$$\dfrac{(-3)+(-5)+7+x}{4} = -1$$

$$\dfrac{-1+x}{4} = -1$$

$$-1+x = -4$$

$$x = -3$$

37. $\dfrac{(-49)+(-72)+(-28)+37}{4} = -\dfrac{112}{4}$

39. $\dfrac{(-10)+(-24)+45+(-16)+(-30)}{5} = \dfrac{-35}{5} = -7$

41. $\dfrac{(-8)+(-3)+12+(-21)+27+(-19)}{6} = \dfrac{-12}{6} = -2$

43. $\dfrac{(-23)+(-34)+(-21)+(-6)+(-15)+0+(-13)}{7} = \dfrac{-112}{7} = -16$

45. $\dfrac{62+(-54)+(-81)+(-34)+(-21)+55+82+87}{8} = \dfrac{96}{8} = 12$

47. $-4(12)+(-3)(-12) = -48+36 = -12$

49. $\dfrac{(-74°)+(-64°)+(-10°)+(-42°)+(-90°)}{5} = \dfrac{-280°}{5} = -56°C$

51. $\dfrac{(-45°)+(-33°)+(-46°)+(-36°)+(-10°)}{5} = \dfrac{-170°}{5} = -34°C$

53. Simplify:

$$[-3+(-6)]^2 - [-8 - 2(-3)]^2$$
$$[-9]^2 - [-8 + 6]^2$$
$$81 - [-2]^2$$
$$81 - 4$$
$$77$$

55. Simplify:

$$\left[46 - 3(-4)^2\right]^3 - \left[-7(1)^3 + (-5)(-8)\right]$$
$$[46 - 3(16)]^3 - [-7(1) + (40)]$$
$$[46 - 48]^3 - [-7 + (40)]$$
$$[-2]^3 - [33]$$
$$-8 - 33$$
$$-41$$

57. Simplify:

$$-15 - \frac{8^2 - 28}{3^2 + 3}$$
$$-15 - \frac{64 - 28}{9 + 3}$$
$$-15 - \frac{36}{12}$$
$$-15 - 3$$
$$-18$$

59. Simplify:

$$\frac{12(8 - 24)}{5^2 - 3^2} \div (-12)$$
$$\frac{12(-16)}{25 - 9} \div (-12)$$
$$\frac{-192}{16} \div (-12)$$
$$(-12) \div (-12)$$
$$1$$

61. Simplify:

$$-8|125 - 321| - 21^2 + 8(-7)$$
$$-8|196| - 441 - 56$$
$$-1568 - 441 - 56$$
$$-2065$$

63. Simplify:

$$-6\left(8^2 - 9^2\right)^2 - (-7)20$$
$$-6(64 - 81)^2 + 140$$
$$-6(17)^2 + 140$$
$$-6(289) + 140$$
$$-1734 + 140$$
$$-1594$$

65. The difference is:

$$\frac{28}{-7} - (-4)(-3)$$
$$(-4) - 12$$
$$-16$$

67. The total price is:

$$\$95 + 15(\$47)$$
$$\$95 + \$705$$
$$\$800$$

69. The loss is:

$$24(-\$3) + 9(\$7)$$
$$-\$72 + \$63$$
$$-\$9$$

71. The averages of the highest and the lowest point of each continent:

Continent	Highest Point	Lowest Point	Average
Africa	+19,340	−512	$\dfrac{19,340-512}{2}=\dfrac{18,828}{2}=9414$
Antarctica	+16,864	−8327	$\dfrac{16,864-8327}{2}=\dfrac{8537}{2}=4268.5$
Asia	+29,028	−1312	$\dfrac{29,028-1312}{2}=\dfrac{27,716}{2}=13,858$
Australia	+7310	−52	$\dfrac{7310-52}{2}=\dfrac{7258}{2}=3629$
Europe	+18,510	−92	$\dfrac{18,510-92}{2}=\dfrac{18,418}{2}=9209$
North America	+20,320	−282	$\dfrac{20,320-282}{2}=\dfrac{20,038}{2}=10,019$
South America	+22,834	−131	$\dfrac{22,834-131}{2}=\dfrac{22,703}{2}=11,351.5$

73. Athens, Bangkok and New Dehli have an average altitude of about 350 ft.
$$\frac{300+0+770}{3}=\frac{1070}{3}=357$$

75. $\$4756 + \$345 - \$212 - \$1218 - \$15 + \$98 - \$450 - \$78 = \$3226$

77. The error is that addition was carried out before multiplication, when multiplication is done prior to addition. The correct answer is:
$$2[3 + 5(-4)] = 2[3 + (-20)] = 2[-17] = -34]$$

79. answers vary

81. answers vary

83. $$\frac{(5-9)^2 + (-6+8)^2 - (14-6)^2}{[3-4(7)+3^3]^2} = \frac{(-4)^2 + (2)^2 - (8)^2}{[3-(28)+27]^2} = \frac{16+4-64}{[2]^2} = \frac{-44}{4} = -11$$

85. answers vary

87. $45y - 18y - 21y = 6y$

89. Solve:
$$14 + w = 57$$
$$14 - 14 + w = 57 - 14$$
$$w = 43$$

Section 4.8
Evaluating Algebraic Expressions and Formulas (Integers)

1. Evaluate:
    $y - 21$
    $14 - 21$
    -7

3. Evaluate:
    $a + b$
    $(-5) + (-7)$
    -12

5. Evaluate:
    $b - c$
    $(-7) - 12$
    -19

7. Evaluate:
    $a - b - c$
    $-5 - (-7) - 12$
    $-5 + 7 - 12$
    -10

9. Evaluate:
    $C = D - R + T$
    $C = 85 - 98 + 11$
    $C = -2$

11. Evaluate:
    $15a - 3b - 5c$
    $15(-7) - 3(-4) - 5(3)$
    $-105 + 12 - 15$
    -108

13. Evaluate:
    $a(3b - 4c) - 5a$
    $-7[3(-4) - 4(3)] - 5(-7)$
    $-7[-12 - 12] + 35$
    $-7[-24] + 35$
    $168 + 35$
    203

15. Evaluate:
    $2a^2 - 4b^2 - 3c^2$
    $2(-7)^2 - 4(-4)^2 - 3(3)^2$
    $2(49) - 4(16) - 3(9)$
    $98 - 64 - 27$
    7

17. 10 is a solution
    $3x - 6 = 24$
    $3(10) - 6 = 24$
    $30 - 6 = 24$
    $24 = 24$

19. Solve:
    $S = v + gt$
    $S = 45 + (-32)(3)$
    $S = 45 - 96$
    $S = -51$

21. Evaluate:

$-4mn - 9p$

$-4(-6)(-8) - 9(-9)$

$-192 + 81$

-111

23. Evaluate:

$mnp - 6m + 3np$

$(-6)(-8)(-9) - 6(-6) + 3(-8)(-9)$

$-432 + 36 + 216$

-180

25. Evaluate:

$m^2 - np^2 + 14p$

$(-6)^2 - (-8)(-9)^2 + 14(-9)$

$36 - (-8)(81) - 126$

$36 + 648 - 126$

558

27. $F = \dfrac{9C + 160}{5} = \dfrac{9(-90°) + 160}{5} = \dfrac{-810° + 160}{5} = \dfrac{-650°}{5} = -130°F$

29. $F = \dfrac{9C + 160}{5} = \dfrac{9(-10°) + 160}{5} = \dfrac{-90° + 160}{5} = \dfrac{70°}{5} = 14°F$

31. 8 is not a solution

$5x - 12 = 2x + 6$

$5(8) - 12 = 2(8) + 6$

$40 - 12 = 16 + 6$

$28 \neq 22$

33. −2 is not a solution

$7a + 12 = 5a + 10$

$7(-2) + 12 = 5(-2) + 10$

$-14 + 12 = -10 + 10$

$-2 \neq 0$

35. −5 is not a solution

$-2b - 13 = -7b - 12$

$-2(-5) - 13 = -7(-5) - 12$

$10 - 13 = 35 - 12$

$-3 \neq 23$

37. Evaluate:

$3abc - xy$

$3(-11)(-12)(25) - (12)(-15)$

$9900 + 180$

$10,080$

39. Evaluate:

$-7abx - 12cy$

$-7(-11)(-12)(12) - 12(25)(-15)$

$11,088 + 4500$

-6588

41. $\dfrac{-cy^2 - 3}{x} = \dfrac{-25(-15)^2 - 3}{12} = \dfrac{-25(225) - 3}{12} = \dfrac{-5625 - 3}{12} = \dfrac{-5628}{12} = -469$

43. Evaluate:

$$\frac{a^2 - 2b^2 - 8xy + 3}{-c - 2y - 9} = \frac{(-11)^2 - 2(-12)^2 - 8(12)(-15) + 3}{-25 - 2(-15) - 9} = \frac{121 - 2(144) - 1440 + 3}{-25 + 30 - 9} =$$

$$\frac{121 - 288 - 1440 + 3}{-25 + 30 - 9} = \frac{1276}{-4} = -319$$

45. $3E = 2035 + (J + P)$ This equation is true.

$$3(17,881) = 2035 + (23,376 + 28,232)$$
$$53.643 = 2035 + 51,608$$
$$53,643 = 53,643$$

47. Evaluate: $3abc = 3(3)(6)(-7) = 54(-7) = -378$

49. $-5cd$ does not always represent a negative number.   It will represent a negative number if c and d are both either positive or both negative.

$-5t^2$ always represents a negative number because t^2 is always positive.

51. Simplify:

$$\frac{3a^2b^2 - 2b^2c^2 - (c - a)}{(a + b)^2 - (b - c)^2}$$

$$\frac{3(3)^2(-1)^2 - 2(-1)^2(-2)^2 - (-2 - 3)}{(3 + (-1))^2 - (-1 - (-2))^2}$$

$$\frac{3(9)(1) - 2(1)(4) + (5)}{(2)^2 - (1)^2}$$

$$\frac{27 - 8 + 5}{4 - 1}$$

$$\frac{24}{3} = 8$$

53. group activity

55. Solve:

$$7t = 126$$
$$\frac{7t}{7} = \frac{126}{7}$$
$$t = 18$$

57. Solve:

$$9y - 7 = 92$$
$$9y - 7 + 7 = 92 + 7$$
$$9y = 99$$
$$\frac{9y}{9} = \frac{99}{9}$$
$$y = 11$$

Section 4.9
Solving Equations (Integers)

1. Solve:
$$a + 11 = -12$$
$$a + 11 - 11 = -12 - 11$$
$$a = -23$$

3. Solve:
$$x - 12 = -22$$
$$x - 12 + 12 = -22 + 12$$
$$x = -10$$

5. Solve:
$$4x = -36$$
$$\frac{4x}{4} = \frac{-36}{4}$$
$$x = -9$$

7. Solve:
$$-12c = -60$$
$$\frac{-12c}{-12} = \frac{-60}{-12}$$
$$c = 5$$

9. Solve:
$$-2x - 8 = 16$$
$$-2x - 8 + 8 = 16 + 8$$
$$-2x = 24$$
$$\frac{-2x}{-2} = \frac{24}{-2}$$
$$x = -12$$

11. Solve:
$$-14 = 37y - 88$$
$$-14 + 88 = 37y - 88 + 88$$
$$74 = 37y$$
$$\frac{74}{37} = \frac{37y}{37}$$
$$2 = y$$

13. Solve:
$$29y + 2 = 205$$
$$29y + 2 - 2 = 205 - 2$$
$$29y = 203$$
$$\frac{29y}{29} = \frac{203}{29}$$
$$y = 7$$

15. Solve:
$$-2x + 5 = -17$$
$$-2x + 5 - 5 = -17 - 5$$
$$-2x = -22$$
$$\frac{-2x}{-2} = \frac{-22}{-2}$$
$$x = 11$$

17. Solve:

$$67w - 34 = -369$$
$$67w - 34 + 34 = -369 + 34$$
$$67w = -335$$
$$\frac{67w}{67} = \frac{-335}{67}$$
$$w = -5$$

19. Solve:

$$\frac{a}{9} - 8 = -2$$
$$\frac{a}{9} - 8 + 8 = -2 + 8$$
$$\frac{a}{9} = 6$$
$$\frac{a}{9} \cdot 9 = 6 \cdot 9$$
$$a = 54$$

21. Solve:

$$3y - 14 = 2y - 14$$
$$3y - 2y - 14 = 2y - 2y - 14$$
$$y - 14 = -14$$
$$y - 14 + 14 = -14 + 14$$
$$y = 0$$

23. Solve:

$$-9x - 26 = 5x + 44$$
$$-9x - 5x - 26 = 5x - 5x + 44$$
$$-14x - 26 = 44$$
$$-14x - 26 + 26 = 44 + 26$$
$$-14x = 70$$
$$\frac{-14x}{-14} = \frac{70}{-14}$$
$$x = -5$$

25. Solve:

$$19 - 2x - 5 = 5x + 7$$
$$14 - 2x = 5x + 7$$
$$14 - 2x + 2x = 5x + 2x + 7$$
$$14 = 7x + 7$$
$$14 - 7 = 7x + 7 - 7$$
$$7 = 7x$$
$$\frac{7}{7} = \frac{7x}{7}$$
$$1 = x$$

27. Solve:

$$-3k + 25 = 1 - 6k$$
$$-3k + 6k + 25 = 1 - 6k + 6k$$
$$3k + 25 = 1$$
$$3k + 25 - 25 = 1 - 25$$
$$3k = -24$$
$$\frac{3k}{3} = \frac{-24}{3}$$
$$k = -8$$

29. Solve:

$$-2(4r - 21) = -78$$
$$-8r + 42 = -78$$
$$-8r + 42 - 42 = -78 - 42$$
$$-8r = -120$$
$$\frac{-8r}{-8} = \frac{-120}{-8}$$
$$r = 15$$

31. The time is:

$$s = v - gt$$
$$-147 = -19 - 32t$$
$$-147 + 19 = -19 + 19 - 32t$$
$$-128 = -32t$$
$$\frac{-128}{-32} = \frac{-32t}{-32}$$
$$4 \sec = t$$

33. The number is:

$$98 + 6x = 266$$
$$98 - 98 + 6x = 266 - 98$$
$$6x = 168$$
$$\frac{6x}{6} = \frac{168}{6}$$
$$x = 28$$

35. The number is:

$$15x - 181 = -61$$
$$15x - 181 + 181 = -61 + 181$$
$$15x = 120$$
$$\frac{15x}{15} = \frac{120}{15}$$
$$x = 8$$

37. Solve:

$$3x + 15 + 21 = 22 + 4x$$
$$3x + 36 = 22 + 4x$$
$$3x - 3x + 36 = 22 + 4x - 3x$$
$$36 = 22 + x$$
$$36 - 22 = 22 - 22 + x$$
$$14 = x$$

39. Solve:

$$40p - 24 - 33p - 77 = 102$$
$$7p - 101 = 102$$
$$7p - 101 + 101 = 102 + 101$$
$$7p = 203$$
$$\frac{7p}{7} = \frac{203}{7}$$
$$p = 29$$

41. Solve:

$$37q + 9 - 8q = 8q - 9 - 87$$
$$29q + 9 = 8q - 96$$
$$29q - 8q + 9 = 8q - 8q - 96$$
$$21q + 9 = -96$$
$$21q + 9 - 9 = -96 - 9$$
$$21q = -105$$
$$\frac{21q}{21} = \frac{-105}{21}$$
$$q = -5$$

43. The completed table is:

F°	122°F	104°F	77°F	50°F	14°F	−13°F	−40°F
C°	50°C	40°C	25°C	10°C	−10°C	−25°C	−40°C

$$5F = 9C + 160°$$
$$5(122°) = 9C + 160°$$
$$610° = 9C + 160°$$
$$450° = 9C$$
$$50° = C$$

$$5F = 9C + 160°$$
$$5(104°) = 9C + 160°$$
$$520° = 9C + 160°$$
$$360° = 9C$$
$$40° = C$$

$$5F = 9C + 160°$$
$$5F = 9(25°) + 160°$$
$$5F = 225° + 160°$$
$$5F = 385°$$
$$F = 77°$$

$$5F = 9C + 160°$$
$$5F = 9(10°) + 160°$$
$$5F = 90° + 160°$$
$$5F = 250°$$
$$F = 50°$$

$$5F = 9C + 160°$$
$$5F = 9(-10°) + 160°$$
$$5F = -90° + 160°$$
$$5F = 70°$$
$$F = 14°$$

$$5F = 9C + 160°$$
$$5(-13°) = 9C + 160°$$
$$-65° = 9C + 160°$$
$$-225° = 9C$$
$$-25° = C$$

$$5F = 9C + 160°$$
$$5(-40°) = 9C + 160°$$
$$-200° = 9C + 160°$$
$$-360° = 9C$$
$$-40° = C$$

45. The value of N is:
$$D = B - NP$$
$$\$575 = \$925 - \$25N$$
$$\$25N = \$350$$
$$N = 14$$

47. The equation that describes this relationship is:
$$7310 = 12{,}558 + 4L$$
$$-5245 = 4L$$
$$-1312 \approx L$$
Asia is the continent whose lowest point fits this description.

49. The equation that describes this relationship is:
$$-35{,}840 = -2000 + 2L$$
$$-33840 = 2L$$
$$-16{,}920 = L$$
The Ionian Basin fits this description.

51. answers vary

53. Solve:
$$3(x-2)+5(x+7)=69$$
$$3x-6+5x+35=69$$
$$8x+29=69$$
$$8x=40$$
$$x=5$$

55. Solve:
$$5(x-4)+2(x-3)-3(x-5)=9$$
$$5x-20+2x-6-3x+15=9$$
$$4x-11=9$$
$$4x=20$$
$$x=5$$

57. The first equation has no solutions, because a false statement is obtained. The second equation has an infinite amount of solutions, because a true statement is obtained. The equations are special because, in the process of solving the them, the variables drop out. Therefore, no variable can be solved for.

59. Solve:
$$\frac{z}{19}=15$$
$$\frac{z}{19}\cdot 19=15\cdot 19$$
$$z=285$$

61. Solve:
$$45y+17=1052$$
$$45y+17-17=1052-17$$
$$45y=1035$$
$$\frac{45y}{45}=\frac{1035}{45}$$
$$y=23$$

Chapter 4
True-False Concept Review

1. False – The opposite of a positive integer is a negative integer.

2. True

3. False – The difference of two negative integers can be positive or negative.

4. True

5. True

6. True

7. False – The quotient of a positive integer and a negative integer is a negative integer.

8. True

9. True

10. False – The product of an even number of negative integer factors is positive.

11. False – Some equations containing integers (coefficients) have solutions that are integers.

12. True

13. True

14. True

15. True

16. True

17. False – The commutative property does not apply to any subtraction problem.

18. True

19. False – The rules for determining the sign of the sum and difference of two integers are different.

20. False - $-5^2 = -25$

Chapter 4
Review

1. The opposite of –22 is 22.

3. If $x = 32$, then $-x = -32$.

5. $-(-8) = 8$

7. $|-87| = 87$

9. $|x| = |-12| = 12$

11. $-43 < -25$

13. $-79 < -32$

15. $-11 < -8$

17. $87 + 43 + (-89) = 41$

19. $x = -123 + (-83) = -206$

21. $-103 - (29) = -103 + (-29) = -132$

23. $97 - (-83) = 97 + 83 = 180$

25. $k = -98 - 98 = -98 + (-98) = -196$

27. $43x - 21x - 45x = 43x - 66x = -23x$

29. $(2a - 5b + 6) + (4a + 3b + 9) = 6a - 2b + 15$

31. $-23(-12) = 276$

33. $(-29)(12) = -348$

35. $x = (-3)(-9)(-21) = -567$

37. $\dfrac{-189}{9} = -21$

39. $\dfrac{-448}{-14} = 32$

41. $12(11a) = 132a$

43. $(-18z)(-6z) = 108z^2$

45. $(2x + 1)(x - 3) = 2x^2 - 6x + x - 3 = 2x^2 - 5x - 3$

47. $\dfrac{-75y^2}{5y} = -15y$

49. $144b^2 \div (-16b) = -9b$

51. Simplify:
$$-9[-12 + 2(3 \cdot 4 - 15)]$$
$$-9[-12 + 2(12 - 15)]$$
$$-9[-12 + 2(-3)]$$
$$-9[-12 - 6]$$
$$-9[-18]$$
$$162$$

53. Simplify:
$$2(12 - 19)^2 + 3(-5 - 18)$$
$$2(-7)^2 + 3(-23)$$
$$2(49) + 3(-23)$$
$$98 - 69$$
$$29$$

55. Simplify:
$$7(3 - 8)^2 - 4(9 - 13)^2 - 3(4)$$
$$7(-5)^2 - 4(-4)^2 - 12$$
$$7(25) - 4(16) - 12$$
$$175 - 64 - 12$$
$$99$$

57. $\dfrac{(-4) + (-7) + 5 + (-13) + (-16)}{5} = \dfrac{-35}{5} = -7$

59. $\dfrac{47 + (-101) + (-341) + (-45) + 153 + 5}{6} = \dfrac{-282}{6} = -47$

61. Evaluate:
$$2a - 3b + 4c$$
$$2(-4) - 3(-3) + 4(9)$$
$$-8 + 9 + 36$$
$$-8 + 45$$
$$33$$

63. Evaluate:
$$-8abc + a$$
$$-8(-4)(-3)(9) + (-4)$$
$$-864 + (-4)$$
$$-868$$

65. Evaluate:

$$y = -8x - 5$$
$$y = -8(-2) - 5$$
$$y = 16 - 5$$
$$y = 11$$

67. Solve:
$$-3(a + 4) = 15$$
$$-3a - 12 = 15$$
$$-3a - 12 + 12 = 15 + 12$$
$$-3a = 27$$
$$\dfrac{-3a}{-3} = \dfrac{27}{-3}$$
$$a = -9$$

69. Solve:
$$2x + 5 - 3x = -2$$
$$-x + 5 = -2$$
$$-x + 5 - 5 = -2 - 5$$
$$-x = -7$$
$$\frac{-x}{-1} = \frac{-7}{-1}$$
$$x = 7$$

Chapter 4
Test

1. $-46 + 34 + (-23) = -69 + 34 = -35$

2. $(-11x)(-3x)(-5) = -165x^2$

3. Evaluate:
$$4a + 5b - 6c$$
$$4(-2) + 5(8) - 6(-11)$$
$$-8 + 40 + 66$$
$$-8 + 106$$
$$98$$

4. Multiply:

$$-3x(2x - 7y - 13)$$
$$-6x^2 + 21xy + 39x$$

5. The opposite of -88 is 88.

6. $87 - 106 = 87 + (-106) = -19$

7. Simplify:
$$-15b + 25b - 37b - 23b$$
$$25b - 75b$$
$$-50b$$

8. Divide:

$$\frac{-595}{35} = -17$$

9. Simplify:
$$2[-3(-2 \cdot 12 - 34) - 4(5)]$$
$$2[-3(-24 - 34) - 20]$$
$$2[-3(-58) - 20]$$
$$2[174 - 20]$$
$$2[154]$$
$$308$$

10. Evaluate:
$$3abc - 2b - 4c$$
$$3(-3)(7)(-4) - 2(7) - 4(-4)$$
$$252 - 2(49) + 16$$
$$252 - 98 + 16$$
$$154 + 16$$
$$170$$

11. Solve:

$$5x + 19 = -36$$

$$5x + 19 - 19 = -36 - 19$$

$$5x = -55$$

$$\frac{5x}{5} = \frac{-55}{5}$$

$$x = -11$$

12. $-7 > -19$

13. $-68 + (-43) + (-34) + 55 = -90$

14. $-44 - (-56) = -44 + 56 = 12$

15. $-3x(4x - 5y + 10) = -12x^2 + 15xy - 30x$

16. Evaluate:

$$\frac{9x - xy^2 - 7y}{3x - 4y} = \frac{9(-6) - (-6)(-4)^2 - 7(-4)}{3(-6) - 4(-4)} =$$

$$\frac{-54 - (-6)(16) + 28}{-18 + 16} = \frac{-54 + 96 + 28}{-18 + 16} = \frac{70}{-2} = -35$$

17. Solve:

$$8a - 12 - 3a = -32 + 7a + 18$$

$$5a - 12 = -14 + 7a$$

$$5a - 5a - 12 = -14 + 7a - 5a$$

$$-12 = -14 + 2a$$

$$-12 + 14 = -14 + 14 + 2a$$

$$2 = 2a$$

$$1 = a$$

18. $(4b^2 - 6b - 11) + (-6b^2 + 3b - 45) = -2b^2 - 3b - 56$

19. Simplify:

$$(6a^2 - 7ab + 5b^2) - (9a^2 + 12ab + 21b^2)$$

$$6a^2 - 7ab + 5b^2 - 9a^2 - 12ab - 21b^2$$

$$-3a^2 - 19ab - 16b^2$$

20. $-5(-7ab)(-4bc) = -140ab^2c$

21. −4 is not a solution

$$5x - 22 - 3x = -50$$
$$5(-4) - 22 - 3(-4) = -50$$
$$-20 - 22 + 12 = -50$$
$$-42 + 12 = -50$$
$$-30 \ne -50$$

22. $|-135| = 135$

23. $\dfrac{-84m^2}{-7m} = 12m$

24. Solve:

$$5d - 32 = -97$$
$$5d - 32 + 32 = -97 + 32$$
$$5d = -65$$
$$\frac{5d}{5} = \frac{-65}{5}$$
$$d = -13$$

25. $-18(-14)(-1)(11) = -2772$

26. Simplify:

$$\frac{4(-5) - (-3)^3 + 5(-6) - 4}{4^2 - 5^2}$$
$$\frac{-20 - (-27) - 30 - 4}{16 - 25}$$
$$\frac{-20 + 27 - 30 - 4}{16 - 25}$$
$$\frac{27}{-9} = -3$$

27. The average is: $\dfrac{-35 + 67 + (-12) + (-43) + 124 + (-89)}{6} = \dfrac{191 - 179}{6} = \dfrac{12}{6} = 2$

28. Solve:

$$7k - 35 - 9k = 3k - 2(k + 4)$$
$$-2k - 35 = 3k - 2k - 8$$
$$-2k - 35 = k - 8$$
$$-27 = 3k$$
$$-9 = k$$

29. The difference between the temperatures is:

$$102°F - (-18°F) = 102°F + (18°F) = 120°F$$

30. The loss is: $5(\$83) + 12(-\$56) = \$415 - \$672 = -\$257$

101

CHAPTER FIVE
FRACTIONS and EQUATIONS

Section 5.1
Opposites, Absolute Value, and Inequalities

1. $\dfrac{4}{7}$

3. $\dfrac{7}{10}$

5. $1\dfrac{1}{3} = \dfrac{4}{3}$

7. $1\dfrac{3}{10} = \dfrac{13}{10}$

9. The proper fractions are: $\dfrac{4}{6}, \dfrac{5}{6}$

11. The proper fractions are: $\dfrac{8}{12}, \dfrac{2}{14}, \dfrac{10}{12}, \dfrac{8}{14}, \dfrac{11}{22}$

13. The proper fractions are: $\dfrac{4}{5}, \dfrac{16}{17}, -\dfrac{99}{100},$

15. The opposite of the listed fractions are: $-\dfrac{13}{19}, \dfrac{17}{14}, \dfrac{12}{17}$

17. $\left|\dfrac{13}{17}\right| = \dfrac{13}{17}$

19. The opposite of the listed fractions are: $\dfrac{16}{19}, \dfrac{14}{3}, -\dfrac{16}{13}$

21. $\left|-\dfrac{14}{27}\right| = \dfrac{14}{27}$

23. $\dfrac{5}{3} = \dfrac{3}{3} + \dfrac{2}{3} = 1\dfrac{2}{3}$

25. $\dfrac{11}{5} = \dfrac{10}{5} + \dfrac{1}{5} = 2\dfrac{1}{5}$

27. $\dfrac{91}{25} = \dfrac{75}{25} + \dfrac{16}{25} = 3\dfrac{16}{25}$

29. $\dfrac{214}{41} = \dfrac{205}{41} + \dfrac{9}{41} = 5\dfrac{9}{41}$

31. $2\dfrac{1}{6} = \dfrac{12}{6} + \dfrac{1}{6} = \dfrac{13}{6}$

33. $9 = \dfrac{9}{1}$

35. $47\dfrac{2}{3} = \dfrac{141}{3} + \dfrac{2}{3} = \dfrac{143}{3}$

37. $106\dfrac{7}{8} = \dfrac{848}{8} + \dfrac{7}{8} = \dfrac{855}{8}$

39. The fraction $\dfrac{0}{\text{any integer but } 0}$ is a proper fraction. Division by zero is undefined.

41. $\left|\dfrac{-7}{16}\right| = +\dfrac{7}{16}$ Absolute values are always non-negative numbers.

43. ___ __ __ __ __ $\downarrow$ __ __ __

45. $\dfrac{25}{25+32} = \dfrac{25}{57}$ are used brick.

47. $\dfrac{843}{2147}$ is spent for housing.

49. $\dfrac{50}{16} = \dfrac{48}{16} + \dfrac{2}{16} = 3\dfrac{2}{16}$;   therefore, 3 is the nearest whole number to $3\dfrac{2}{16}$

51. $7\dfrac{31}{352} = \dfrac{2464}{352} + \dfrac{31}{352} = \dfrac{2495}{352}$;   therefore, 2495 sections are needed.

53. The probability that a 56-year old man is balding is $\dfrac{1}{2}$.

55. The 70-79 year old group has the highest probability ($\dfrac{3}{4}$) of being bald.

57. The four possible outcomes of two people flipping one coin each are:

HH with a probability of $\dfrac{1}{4}$ TT with a probability of $\dfrac{1}{4}$

HT with a probability of $\dfrac{1}{4}$ TH with a probability of $\dfrac{1}{4}$

Therefore, the probability of getting one head and one tail is $\dfrac{1}{4} + \dfrac{1}{4} = \dfrac{2}{4} = \dfrac{1}{2}$

59. answers vary

61. answers vary

63. $13 = \dfrac{13}{1} = \dfrac{13}{1} \cdot \dfrac{9}{9} = \dfrac{117}{9}$ $13 = \dfrac{13}{1} = \dfrac{13}{1} \cdot \dfrac{117}{117} = \dfrac{1521}{117}$

65. Some examples are: $\dfrac{47}{6} = 7\dfrac{5}{6} = 6\dfrac{11}{6} = 5\dfrac{17}{6} = 4\dfrac{23}{6} = 3\dfrac{29}{6}$

67. answers vary

69. answers vary

71. $|-29| = 29$

73. $72 > -73$

Section 5.2
Multiples, Factors and Prime Factorization

1. The first five multiples of 2 are: 2, 4, 6, 8, 10

3. The first five multiples of 5 are: 5, 10, 15, 20, 25

5. The first five multiples of 11 are: 11, 22, 33, 44, 55

7. The first five multiples of 25 are: 25, 50, 75, 100, 125

9. The first five multiples of 135 are: 135, 270, 405, 540, 675

11. 38 is not a multiple of 4

13. 48 is a multiple of 4: $4 \cdot 12 = 48$

15. 116 is a multiple of 4: $4 \cdot 29 = 116$

17. 290 is not a multiple of 4

19. The multiples of 8 from 72 to 112 are: 72, 80, 88, 96, 104, 112

21. 30 is divisible by 2, 3, 5 and 10

23. 135 is divisible by 3 and 5

25. 370 is divisible by 2, 5 and 10

27. 7880 is divisible by 2, 5 and 10

29. 11,305 is divisible by 5

31. The factors of 12 are 1, 2, 3, 4, 6 and 12

33. The factors of 13 are 1 and 13

35. The factors of 36 are 1, 2, 3, 4, 6, 9, 12, 18 and 36

37. The factors of 111 are 1, 3, 37 and 111

39. The factors of 405 are 1, 3, 5, 9, 15, 27, 45, 81, 135 and 405

41. 29 is prime

43. 57 is composite: $3 \cdot 19$

45. 497 is composite: $7 \cdot 71$

47. 797 is prime

49. $63 = 3^2 \cdot 7$

51. $44 = 2^2 \cdot 11$

53. $121 = 11^2$

55. $190 = 2 \cdot 5 \cdot 19$

57. $1536 = 2^9 \cdot 3$

59. 3711 is a multiple of 3
$$3711 = 3(1237)$$
3741 is a multiple of 3
$$3741 = 3(1247)$$
3771 is a multiple of 3
$$3771 = 3(1257)$$

61. Six is a <u>factor</u> of 66.

63. The student should work problems 4, 8, 12, 16, 20, 24, 28, 32, 36, 40, 44, 48 and 52.

65. The outlet does get the discount price for 176 chairs because 176 is a multiple of 16.
$$176 = 16 \cdot 11$$

67. The first year in the 21st century that is a multiple of 75 is 2025: $2025 = 27 \cdot 75$

69. The next prime number year after 1997 is 1999.

71. Jorge bets the smallest four digit prime number, which is 1009.

73. 9 is a <u>factor</u> of 126 because 9 can be multiplied by 14 to obtain 126.
9 is a <u>divisor</u> of 126 because 126 can be divided by 9 fourteen times.

75. The number 2 is the only even prime number because all of the other even numbers are divisible by 2, as well as by themselves and 1.

77. There are 1333 multiples of 3 between 1000 and 5000.
$$5000 - 1000 = 4000, \qquad 4000 \div 3 = 1333.\overline{3}$$

79. A number that is divisible by 6 if it is divisible by both 2 and 3.
A number that is divisible by 15 if it is divisible by both 3 and 5.

81. The number of multiples of 6 between 1000 and 5000 is 666.
$$5000 - 1000 = 4000, \qquad 4000 \div 6 = 666.\overline{6}$$

83. $(-2)(111)(18) = -3996$

85. $(-1176) \div 14 = -84$

87. $468xy^2 \div -6xy = -78y$

Section 5.3
Multiplying and Dividing Fractions

1. $\dfrac{16}{20} = \dfrac{2 \cdot 2 \cdot 2 \cdot 2}{2 \cdot 2 \cdot 5} = 1 \cdot 1 \cdot \dfrac{2 \cdot 2}{5} = \dfrac{4}{5}$

3. $\dfrac{44}{88} = \dfrac{2 \cdot 2 \cdot 11}{2 \cdot 2 \cdot 2 \cdot 11} = 1 \cdot 1 \cdot 1 \cdot \dfrac{1}{2} = \dfrac{1}{2}$

5. $\dfrac{28}{40} = \dfrac{2 \cdot 2 \cdot 7}{2 \cdot 2 \cdot 2 \cdot 5} = 1 \cdot 1 \cdot \dfrac{7}{2 \cdot 5} = \dfrac{7}{10}$

7. $\dfrac{45}{75} = \dfrac{3 \cdot 3 \cdot 5}{3 \cdot 5 \cdot 5} = 1 \cdot 1 \cdot \dfrac{3}{5} = \dfrac{3}{5}$

9. $\dfrac{56b}{4b} = \dfrac{2 \cdot 2 \cdot 2 \cdot 7 \cdot b}{2 \cdot 2 \cdot b} = 1 \cdot 1 \cdot 1 \cdot \dfrac{2 \cdot 7}{1} = 14$

11. $\dfrac{64w}{72} = \dfrac{2 \cdot 2 \cdot 2 \cdot 2 \cdot 2 \cdot 2 \cdot w}{2 \cdot 2 \cdot 2 \cdot 3 \cdot 3} = 1 \cdot 1 \cdot 1 \cdot \dfrac{2 \cdot 2 \cdot 2 \cdot w}{3 \cdot 3} = \dfrac{8w}{9}$

13. $\dfrac{546xy^2}{910x} = \dfrac{2\cdot3\cdot7\cdot13\cdot x\cdot y\cdot y}{2\cdot5\cdot7\cdot13\cdot x} = 1\cdot1\cdot1\cdot1\cdot\dfrac{3\cdot y\cdot y}{5} = \dfrac{3y^2}{5}$

15. $\dfrac{2}{3}\cdot\dfrac{2}{2} = \dfrac{4}{6}$ $\dfrac{2}{3}\cdot\dfrac{3}{3} = \dfrac{6}{9}$ $\dfrac{2}{3}\cdot\dfrac{4}{4} = \dfrac{8}{12}$ $\dfrac{2}{3}\cdot\dfrac{5}{5} = \dfrac{10}{15}$

17. $\dfrac{5c}{4}\cdot\dfrac{2}{2} = \dfrac{10c}{8}$ $\dfrac{5c}{4}\cdot\dfrac{3}{3} = \dfrac{15c}{12}$ $\dfrac{5c}{4}\cdot\dfrac{4}{4} = \dfrac{20c}{16}$ $\dfrac{5c}{4}\cdot\dfrac{5}{5} = \dfrac{25c}{20}$

19. $\dfrac{7}{16}\cdot\dfrac{2}{2} = \dfrac{14}{32}$ $\dfrac{7}{16}\cdot\dfrac{3}{3} = \dfrac{21}{48}$ $\dfrac{7}{16}\cdot\dfrac{4}{4} = \dfrac{28}{64}$ $\dfrac{7}{16}\cdot\dfrac{5}{5} = \dfrac{35}{80}$

21. $\dfrac{8}{15}\cdot\dfrac{2}{2} = \dfrac{16}{30}$ $\dfrac{8}{15}\cdot\dfrac{3}{3} = \dfrac{24}{45}$ $\dfrac{8}{15}\cdot\dfrac{4}{4} = \dfrac{32}{60}$ $\dfrac{8}{15}\cdot\dfrac{5}{5} = \dfrac{40}{75}$

23. $\dfrac{2}{3}\cdot\dfrac{8}{8} = \dfrac{16}{24}$ 25 $\dfrac{6}{7}\cdot\dfrac{4}{4} = \dfrac{24}{28}$

27. $\dfrac{11x}{13}\cdot\dfrac{4}{4} = \dfrac{44x}{52}$ 29. $\dfrac{6x^2}{7}\cdot\dfrac{12}{12} = \dfrac{72x^2}{84}$

31. $\dfrac{5}{8}\cdot\dfrac{4}{7} = \dfrac{5}{2\cdot2\cdot2}\cdot\dfrac{2\cdot2}{7} = \dfrac{5}{2\cdot7} = \dfrac{5}{14}$ 33. $\dfrac{5}{8}\cdot\dfrac{4}{15} = \dfrac{5}{2\cdot2\cdot2}\cdot\dfrac{2\cdot2}{3\cdot5} = \dfrac{1}{6}$

35. $\dfrac{6}{7}\cdot\dfrac{14}{15} = \dfrac{2\cdot3}{7}\cdot\dfrac{2\cdot7}{3\cdot5} = \dfrac{2\cdot2}{5} = \dfrac{4}{5}$ 37. $\dfrac{9}{12}\left(\dfrac{10}{15}\right) = \dfrac{3\cdot3}{2\cdot2\cdot3}\left(\dfrac{2\cdot5}{3\cdot5}\right) = \dfrac{1}{2}$

39. $\left(-\dfrac{3}{5}\right)\left(\dfrac{5}{2}\right)\left(-\dfrac{2}{3}\right) = 1$

41. $\left(2\dfrac{5}{8}\right)\left(\dfrac{4}{21}\right) = \left(\dfrac{21}{8}\right)\left(\dfrac{4}{21}\right) = \left(\dfrac{3\cdot7}{2\cdot2\cdot2}\right)\left(\dfrac{2\cdot2}{3\cdot7}\right) = \dfrac{1}{2}$

43. $\left(3\dfrac{1}{8}\right)\left(4\dfrac{3}{5}\right)(3) = \left(\dfrac{25}{8}\right)\left(\dfrac{23}{5}\right)\left(\dfrac{3}{1}\right) = \left(\dfrac{5\cdot5}{2\cdot2\cdot2}\right)\left(\dfrac{23}{5}\right)\left(\dfrac{3}{1}\right) = \dfrac{5\cdot23\cdot3}{2\cdot2\cdot2} = \dfrac{345}{8} = 43\dfrac{1}{8}$

45. $\left(\dfrac{56a}{65}\right)\left(\dfrac{39b}{48}\right)\left(\dfrac{18b}{25}\right) = \left(\dfrac{2\cdot2\cdot2\cdot7a}{5\cdot13}\right)\left(\dfrac{3\cdot13b}{2\cdot2\cdot2\cdot2\cdot3}\right)\left(\dfrac{2\cdot3\cdot3b}{5\cdot5}\right) = \dfrac{7\cdot3\cdot3ab^2}{5\cdot5\cdot5} = \dfrac{63ab^2}{125}$

47. $\dfrac{2}{5}\div\dfrac{3}{5} = \dfrac{2}{5}\cdot\dfrac{5}{3} = \dfrac{2}{3}$

49. $\dfrac{8}{9} \div \left(-\dfrac{7}{18}\right) = \dfrac{8}{9} \cdot \left(-\dfrac{18}{7}\right) = \dfrac{2\cdot2\cdot2}{3\cdot3} \cdot \left(-\dfrac{2\cdot3\cdot3}{7}\right) = -\dfrac{2\cdot2\cdot2\cdot2}{7} = -\dfrac{16}{7}$

51. $\dfrac{8}{15} \div \dfrac{16}{5} = \dfrac{8}{15} \cdot \dfrac{5}{16} = \dfrac{8}{3\cdot5} \cdot \dfrac{5}{2\cdot8} = \dfrac{1}{2\cdot3} = \dfrac{1}{6}$

53. $\left(-\dfrac{5}{18}\right) \div \left(\dfrac{10}{27}\right) = \left(-\dfrac{5}{18}\right) \cdot \left(\dfrac{27}{10}\right) = \left(-\dfrac{5}{2\cdot9}\right) \cdot \left(\dfrac{3\cdot9}{2\cdot5}\right) = -\dfrac{3}{2\cdot2} = -\dfrac{3}{4}$

55. $\left(-\dfrac{20xy}{21}\right) \div \left(\dfrac{9}{10}\right) = \left(-\dfrac{20xy}{21}\right) \cdot \left(\dfrac{10}{9}\right) = \left(-\dfrac{2\cdot2\cdot5}{3\cdot7}xy\right) \cdot \left(\dfrac{2\cdot5}{3\cdot3}\right) = -\dfrac{200}{189}xy$

57. $\left(\dfrac{5a}{72}\right) \div \left(\dfrac{2a}{25}\right) = \left(\dfrac{5a}{72}\right)\left(\dfrac{25}{2a}\right) = \dfrac{125}{144}$

59. $\dfrac{2}{3} \cdot \dfrac{5}{16} = \dfrac{2}{3} \cdot \dfrac{5}{2\cdot8} = \dfrac{5}{24}$

61. To correct the problem $\dfrac{3}{8} \cdot \dfrac{4}{5} = \dfrac{15}{32}$, change the multiplication sign to division:

$$\dfrac{3}{8} \div \dfrac{4}{5} = \dfrac{3}{8} \cdot \dfrac{5}{4} = \dfrac{15}{32}$$

63. $\dfrac{243a}{1000} \cdot \dfrac{25}{81} \cdot \dfrac{8}{9} = \dfrac{3\cdot3\cdot3\cdot3\cdot3\cdot a}{2\cdot2\cdot2\cdot5\cdot5\cdot5} \cdot \dfrac{5\cdot5}{3\cdot3\cdot3\cdot3} \cdot \dfrac{2\cdot2\cdot2}{3\cdot3} = \dfrac{a}{15}$

65. $\left(-\dfrac{16s^2t}{81}\right) \div \left(-\dfrac{8st}{108}\right) = \left(-\dfrac{16s^2t}{81}\right) \cdot \left(-\dfrac{108}{8st}\right) = \left(-\dfrac{2\cdot2\cdot2\cdot2s\cdot ts}{3\cdot3\cdot3\cdot3}\right) \cdot \left(-\dfrac{2\cdot2\cdot3\cdot3\cdot3}{2\cdot2\cdot2st}\right) = \dfrac{8s}{3}$

67. 4 is a solution

$$\dfrac{11}{6}x = \dfrac{22}{3}$$

$$\dfrac{11}{6} \cdot 4 = \dfrac{22}{3}$$

$$\dfrac{11}{3} \cdot 2 = \dfrac{22}{3}$$

$$\dfrac{2}{3} = \dfrac{22}{3}$$

69. $\dfrac{1}{2} \cdot \dfrac{12}{12} = \dfrac{12}{24}$ $\dfrac{2}{3} \cdot \dfrac{8}{8} = \dfrac{16}{24}$ $\dfrac{1}{6} \cdot \dfrac{4}{4} = \dfrac{4}{24}$ $\dfrac{5}{8} \cdot \dfrac{3}{3} = \dfrac{15}{24}$

71. $\left(62\dfrac{1}{2}\right)\left(2\dfrac{13}{20}\right)(5800) = \left(\dfrac{125}{2}\right)\left(\dfrac{53}{20}\right)\left(\dfrac{5800}{1}\right) = 960{,}625$ lb

73. $\dfrac{12}{28} = \dfrac{3}{7}$

75. $\dfrac{23}{8} \div \dfrac{1}{4} = \dfrac{23}{8} \cdot \dfrac{4}{1} = \dfrac{23}{2} = 11\dfrac{1}{2}$ omelets

77. $\dfrac{1}{8} \cdot 3\dfrac{1}{2} \cdot 5 = \dfrac{1}{8} \cdot \dfrac{7}{2} \cdot \dfrac{5}{1} = \dfrac{35}{16} = 2\dfrac{3}{16}$ in^3

79. $\left(8\dfrac{3}{4}\,\text{in}\right)\left(1100\,\text{ft}^2\right) = \left(\dfrac{35}{4}\,\text{in}\right)\left(1100\,\text{ft}^2\right)\left(\dfrac{144\,\text{in}^2}{\text{ft}^2}\right) = 1{,}386{,}000\,\text{in}^3$

81. The probability of drawing a black card is: $\dfrac{26}{52} = \dfrac{1}{2}$ and the probability of drawing a

seven is: $\dfrac{4}{52} = \dfrac{1}{13}$ Then the probability of drawing a black seven is: $\dfrac{1}{2} \cdot \dfrac{1}{13} = \dfrac{1}{26}$

83. The probability that the student is male is: $\dfrac{18}{30} = \dfrac{3}{5}$.

The probability that the student is brown-eyed is: $\dfrac{25}{30} = \dfrac{5}{6}$.

The probability that the student is a brown-eyed male is: $\dfrac{3}{5} \cdot \dfrac{5}{6} = \dfrac{1}{2}$

85. answers vary

87. $\dfrac{15}{25} = \dfrac{3 \cdot 5}{5 \cdot 5} = \dfrac{3}{5}$ $\quad$ $\dfrac{3}{5} \cdot \dfrac{2}{2} = \dfrac{6}{10}$ $\quad$ $\dfrac{3}{5} \cdot \dfrac{4}{4} = \dfrac{12}{20}$ $\quad$ $\dfrac{3}{5} \cdot \dfrac{6}{6} = \dfrac{18}{30}$ $\quad$ $\dfrac{3}{5} \cdot \dfrac{12}{12} = \dfrac{36}{60}$

89. answers vary $\qquad$ 91. answers vary $\qquad$ 93. $\dfrac{11}{10}$

95. Evaluate: $\qquad$ 97. Evaluate: $\dfrac{4x-12}{a} = \dfrac{4(-6)-12}{-18} = \dfrac{-24-12}{-18} = \dfrac{-36}{-18} = 2$

$-6abc - 5a$

$-6(-3)(-3)(-3) - 5(-3)$

$162 + 15$

177

109

Section 5.4
Conversion of Units within a System

1. 2 weeks(7 day/wk) = 14 days

3. 2 years(12 mo) = 24 months

5. 2 feet(12 in/ft) = 24 inches

7. 1 mile = 5280 feet

9. 2 meters(100 cm/m) = 200 cm

11. $1 yd^2 = 9 ft^2$

13. 80 oz ÷ 16 oz/lb = 5 lb

15. 15 cm(10 mm/cm) = 150 mm

17. 40 pt ÷ 8 pt/gal = 5 gal

19. 6000 g ÷ 1000 g/kg = 6 kg

21. 10 kg(1000 g/kg) = 10,000 g

23. 108 in ÷ 36 in/yd = 3 yd

25. 5 m(100 cm/m) = 500 cm

27. $\dfrac{144\ lb}{1\ ft} \cdot \dfrac{ft}{12\ in} = \dfrac{12\ lb}{in}$

29. $1000\ mm\left(\dfrac{m}{1000\ mm}\right) = 1\ m$

31. $32\ in\left(\dfrac{1\ ft}{12\ in}\right) = \dfrac{8}{3}\ ft = 2\dfrac{2}{3}\ ft$

33. $3\ yd\ 1\ ft = (3\ yd)\dfrac{3\ ft}{yd} + 1\ ft = 9\ ft + 1\ ft = 10\ ft\left(\dfrac{12\ in}{ft}\right) = 120\ in$

35. $3000\ lb\left(\dfrac{1\ ton}{2000\ lb}\right) = \dfrac{3}{2}\ ton = 1\dfrac{1}{2}\ ton$

37. $10,080\ min\left(\dfrac{1\ hr}{60\ min}\right)\left(\dfrac{1\ day}{24\ hr}\right) = 7\ days$

39. $6\dfrac{1}{2}\ tons = \left(\dfrac{13}{2}\ tons\right)\left[\dfrac{2000\ lb}{1\ ton}\right] = 13,000\ lb$

41. $\left(\dfrac{45\ mi}{hr}\right)\left(\dfrac{5280\ ft}{mi}\right)\left(\dfrac{1\ hr}{3600\ sec}\right) = \dfrac{66\ ft}{1\ sec}$

43. $\left(\dfrac{9\ tons}{1\ ft}\right)\left(\dfrac{2000\ lb}{1\ ton}\right)\left(\dfrac{1\ ft}{12\ in}\right) = \dfrac{1500\ lb}{1\ in}$

45. $\left(\dfrac{180\text{ km}}{1\text{ hr}}\right)\left(\dfrac{1\text{ hr}}{3600\text{ sec}}\right)\left(\dfrac{1000\text{ m}}{1\text{ km}}\right) = \dfrac{50\text{ m}}{1\text{ sec}}$

47. $\left(\dfrac{24\text{ g}}{1\text{ m}^2}\right)\left(\dfrac{1\text{ kg}}{1000\text{ g}}\right)\left(\dfrac{1000\text{ m}}{\text{km}}\right)^2 = \dfrac{24{,}000\text{ kg}}{1\text{ km}^2}$

49. $\left(\dfrac{30\text{ tons}}{1\text{ day}}\right)\left(\dfrac{2000\text{ lb}}{1\text{ ton}}\right)\left(\dfrac{1\text{ day}}{24\text{ hr}}\right)\left(\dfrac{1\text{ hr}}{60\text{ min}}\right) = \dfrac{125\text{ lb}}{3\text{ min}} = 41\dfrac{2}{3}\text{ lb/min}$

51. $\left(\dfrac{90\text{ words}}{1\text{ min}}\right)\left(\dfrac{1\text{ min}}{60\text{ sec}}\right) = \dfrac{3\text{ words}}{12\text{ sec}} = 1\dfrac{1}{2}\text{ words/sec}$

53. $\left(\dfrac{0.03\text{ g}}{1}\right)\left(\dfrac{1000\text{ mg}}{1\text{ g}}\right)\left(\dfrac{1\text{ cc}}{2\text{ mg}}\right) = 15\text{ cc}$

55. $\left(\dfrac{12\text{ cents}}{\text{mg}}\right)\left(\dfrac{1{,}000{,}000\text{ mg}}{\text{kg}}\right)\left(\dfrac{2\text{ kg}}{1}\right)\left(\dfrac{\$1}{100\text{ cents}}\right) = \$240{,}000$

57. $3\text{ in}\left(\dfrac{1\text{ min}}{\frac{5}{8}\text{ in}}\right) = \dfrac{24}{5}\text{ min} = 4\dfrac{4}{5}\text{ min}$

59. $\left(\dfrac{64\text{ lb}}{1\text{ ft}^3}\right)\left(\dfrac{1\text{ ft}^3}{1728\text{ in}^3}\right) = 0.\overline{037}\text{ lb} = \dfrac{1}{27}\text{ lb}$

61. $\left(\dfrac{100\text{ kg}}{1\text{ m}^3}\right)\left(\dfrac{1\text{ m}^3}{1000^3\text{ cm}^3}\right)\left(\dfrac{300\text{ cm}^3}{1\text{ bottle}}\right)\cdot\dfrac{1000\text{ g}}{1\text{ kg}} = \dfrac{30{,}000\text{ g}}{1000\text{ bottle}} = 30\text{ g/bottle}$

The price of one bottle is $\dfrac{\$0.20}{\text{g}}\cdot\dfrac{30\text{ g}}{\text{bottle}} = \$6.00/\text{bottle}$

63. $\left(\dfrac{12{,}000\text{ sec}}{1}\right)\left(\dfrac{1\text{ hr}}{3600\text{ sec}}\right) = \dfrac{10}{3}\text{ hr} = 3\dfrac{1}{3}\text{ hr}$

65. $\left(\dfrac{\$5}{1\text{ g}}\right)\left(\dfrac{100\text{ cents}}{\$1}\right)\left(\dfrac{1\text{ g}}{100\text{ cg}}\right) = \dfrac{5\text{ cents}}{1\text{ cg}};$

therefore, the profit is $\left(7\dfrac{1}{2}-5\right)\text{ cents/cg} = 2\dfrac{1}{2}\text{ cents/cg}\left(\dfrac{100{,}000\text{ cg}}{1\text{ kg}}\right) = \$2500/\text{kg}$

67. group activity

69. $188 - (-124) = 188 + (+124) = 312$

71. -14 is a solution
$$x^2 - 3x = 238$$
$$(-14)^2 - 3(-14) = 238$$
$$196 + 42 = 238$$
$$238 = 238$$

Section 5.5
Adding and Subtracting Rational Numbers

1. The LCM of 7 and 28 is 28

3. The LCM of 2, 3 and 5 is 30

5. The LCM of 2, 3 and 6 is 6

7. The LCM of 2, 14 and 21 is 42

9. The LCM of 4, 10, 12, 15 and 20 is 60.
$4 = 2^2$ $10 = 2 \cdot 5$ $12 = 2^2 \cdot 3$ $15 = 3 \cdot 5$ $20 = 2^2 \cdot 5$
LCM $= 2^2 \cdot 3 \cdot 5 = 60$

11. The LCM of 48, 96 and 120 is 480.
$48 = 2^4 \cdot 3$ $96 = 2^5 \cdot 3$ $120 = 2^3 \cdot 3 \cdot 5$
LCM $= 2^5 \cdot 3 \cdot 5 = 480$

13. The LCM of $12x, 17x^2, 51$ and 68 is $204x^2$
$12x = 2^2 \cdot 3 \cdot x$ $17x^2 = 17x^2$ $51 = 3 \cdot 17$ $68 = 2^2 \cdot 17$
LCM $= 2^2 \cdot 3 \cdot 17 \cdot x^2 = 204x^2$

15. $\dfrac{19}{21} < \dfrac{20}{21}$

17. $-\dfrac{3}{13} > -\dfrac{5}{13}$

19. $-\dfrac{4}{21} < \dfrac{3}{7}$

21. $\dfrac{13}{15} > \dfrac{21}{25}$

$\dfrac{13}{15} \cdot \dfrac{5}{5} > \dfrac{21}{25} \cdot \dfrac{3}{3}$

$\dfrac{65}{75} > \dfrac{63}{75}$

23. $\dfrac{20}{33} < \dfrac{8}{11}$

$\dfrac{20}{33} < \dfrac{8}{11} \cdot \dfrac{3}{3}$

$\dfrac{20}{33} < \dfrac{24}{33}$

25. $-\dfrac{7}{30} < -\dfrac{2}{9}$

$-\dfrac{7}{30} \cdot \dfrac{3}{3} < -\dfrac{2}{9} \cdot \dfrac{10}{10}$

$-\dfrac{21}{90} < -\dfrac{20}{90}$

27. $\dfrac{2}{15} + \dfrac{4}{15} + \dfrac{3}{15} = \dfrac{9}{15} = \dfrac{3}{5}$

29. $\dfrac{3}{8}x + \dfrac{5}{8}x = \dfrac{8}{8}x = 1x = x$

31. $\dfrac{5}{48} + \left(-\dfrac{7}{48}\right) + \dfrac{3}{48} = \dfrac{1}{48}$

33. $\dfrac{1}{8} + \dfrac{7}{24} = \dfrac{1}{8} \cdot \dfrac{3}{3} + \dfrac{7}{24} = \dfrac{3}{24} + \dfrac{7}{24} = \dfrac{10}{24} = \dfrac{5}{12}$

35. $\dfrac{1}{2}y + \dfrac{1}{3}y + \dfrac{1}{6}y = \dfrac{1}{2} \cdot \dfrac{3}{3}y + \dfrac{1}{3} \cdot \dfrac{2}{2}y + \dfrac{1}{6}y = \dfrac{3}{6}y + \dfrac{2}{6}y + \dfrac{1}{6}y = \dfrac{6}{6}y = y$

37. $\dfrac{5}{14} + \left(-\dfrac{1}{7}\right) + \left(-\dfrac{2}{7}\right) = \dfrac{5}{14} + \left(-\dfrac{1}{7} \cdot \dfrac{2}{2}\right) + \left(-\dfrac{2}{7} \cdot \dfrac{2}{2}\right) = \dfrac{5}{14} + \left(-\dfrac{2}{14}\right) + \left(-\dfrac{4}{14}\right) = -\dfrac{1}{14}$

39. $\dfrac{7}{16} + \dfrac{3}{20} + \dfrac{1}{5} = \dfrac{7}{16} \cdot \dfrac{5}{5} + \dfrac{3}{20} \cdot \dfrac{4}{4} + \dfrac{1}{5} \cdot \dfrac{16}{16} = \dfrac{35}{80} + \dfrac{12}{80} + \dfrac{16}{80} = \dfrac{63}{80}$

41. $\dfrac{3}{10}w + \dfrac{7}{20}w + \dfrac{11}{30}w = \dfrac{3}{10} \cdot \dfrac{6}{6}w + \dfrac{7}{20} \cdot \dfrac{3}{3}w + \dfrac{11}{30} \cdot \dfrac{2}{2}w = \dfrac{18}{60}w + \dfrac{21}{60}w + \dfrac{22}{60}w = \dfrac{61}{60}w$

43. $\dfrac{5}{6}x + \left(-\dfrac{7}{8}\right) + \dfrac{3}{4} + \left(-\dfrac{1}{2}x\right) = \dfrac{5}{6} \cdot \dfrac{4}{4}x + \left(-\dfrac{7}{8} \cdot \dfrac{3}{3}\right) + \dfrac{3}{4} \cdot \dfrac{6}{6} + \left(-\dfrac{1}{2} \cdot \dfrac{12}{12}x\right) =$

$\dfrac{20x - 21 + 18 - 12x}{24} = \dfrac{8x + 3}{24} = \dfrac{1}{3}x + \dfrac{1}{8}$

45. $\dfrac{29xy}{200} + \dfrac{7xy}{50} + \left(-\dfrac{9xy}{25}\right) = \dfrac{29xy}{200} + \dfrac{4}{4} \cdot \dfrac{7xy}{50} + \left(-\dfrac{8}{8} \cdot \dfrac{9xy}{25}\right) =$

$\dfrac{29xy}{200} + \dfrac{28xy}{200} + \left(-\dfrac{72xy}{200}\right) = -\dfrac{15xy}{200} = -\dfrac{3xy}{40}$

47. $\dfrac{3}{8} - \dfrac{1}{8} = \dfrac{2}{8} = \dfrac{1}{4}$

49. $\dfrac{17}{30} - \dfrac{7}{30} = \dfrac{10}{30} = \dfrac{1}{3}$

51. Subtract:

$$\frac{5}{18}-\frac{2}{9}$$

$$\frac{5}{18}-\frac{2}{9}\cdot\frac{2}{2}$$

$$\frac{5}{18}-\frac{4}{18}$$

$$\frac{1}{18}$$

53. Subtract:

$$\frac{5}{6}-\left(-\frac{1}{2}\right)$$

$$\frac{5}{6}+\left(+\frac{1}{2}\cdot\frac{3}{3}\right)$$

$$\frac{5}{6}+\frac{3}{6}$$

$$\frac{8}{6}=\frac{4}{3}$$

55. Subtract:

$$\frac{8}{9}-\frac{5}{6}$$

$$\frac{8}{9}\cdot\frac{2}{2}-\frac{5}{6}\cdot\frac{3}{3}$$

$$\frac{16}{18}-\frac{15}{18}$$

$$\frac{1}{18}$$

57. Subtract:

$$-\frac{9}{10}-\frac{3}{4}$$

$$-\frac{9}{10}\cdot\frac{2}{2}-\frac{3}{4}\cdot\frac{5}{5}$$

$$-\frac{18}{20}-\frac{15}{20}$$

$$-\frac{33}{20}$$

59. Subtract:

$$\frac{33}{35}st-\frac{17}{28}st$$

$$\frac{33}{35}\cdot\frac{4}{4}st-\frac{17}{28}\cdot\frac{5}{5}st$$

$$\frac{132}{140}st-\frac{85}{140}st$$

$$\frac{47}{140}st$$

61. Subtract:

$$\frac{3}{10}x^2-\frac{1}{5}x^2-\frac{4}{15}x^2$$

$$\frac{3}{10}\cdot\frac{3}{3}x^2-\frac{1}{5}\cdot\frac{6}{6}x^2-\frac{4}{15}\cdot\frac{2}{2}x^2$$

$$\frac{9}{30}x^2-\frac{6}{30}x^2-\frac{8}{30}x^2$$

$$-\frac{5}{30}x^2=-\frac{1}{6}x^2$$

63. $\quad 1\frac{3}{7}+5\frac{2}{7}=6\frac{5}{7}$

65. $\quad 13\frac{6}{7}-8\frac{4}{7}=5\frac{2}{7}$

67. Add:

$$8\frac{5}{12}+6\frac{1}{6}$$

$$8+\frac{5}{12}+6+\frac{1}{6}\cdot\frac{2}{2}$$

$$8+\frac{5}{12}+6+\frac{2}{12}$$

$$14\frac{7}{12}$$

69. $10\frac{2}{5}-3\frac{1}{10}=10+\left(\frac{2}{5}\cdot\frac{2}{2}\right)-3\frac{1}{10}=10\frac{4}{10}-3\frac{1}{10}=7\frac{3}{10}$

71. $1\frac{1}{3}+8\frac{1}{3}-5\frac{1}{6}=1\frac{1}{3}+8\frac{1}{3}-5\frac{1}{6}=9\frac{2}{3}-5\frac{1}{6}=9+\left(\frac{2}{3}\cdot\frac{2}{2}\right)-5\frac{1}{6}=9\frac{4}{6}-5\frac{1}{6}=4\frac{3}{6}=4\frac{1}{2}$

73 $2\frac{2}{5}+7\frac{1}{6}+1\frac{4}{15}=10+\left(\frac{2}{5}\cdot\frac{6}{6}+\frac{1}{6}\cdot\frac{5}{5}+\frac{4}{15}\cdot\frac{2}{2}\right)=10+\left(\frac{12}{30}+\frac{5}{30}+\frac{8}{30}\right)=10\frac{25}{30}=10\frac{5}{6}$

75. $18\frac{7}{12}-9\frac{1}{4}=9+\frac{7}{12}-\frac{1}{4}=9+\frac{7}{12}-\frac{1}{4}\cdot\frac{3}{3}=9+\frac{7}{12}-\frac{3}{12}=9\frac{4}{12}=9\frac{1}{3}$

77. $11\frac{1}{5}+3\frac{7}{10}+12\frac{1}{2}=26+\left(\frac{1}{5}\cdot\frac{2}{2}+\frac{7}{10}+\frac{1}{2}\cdot\frac{5}{5}\right)=26+\left(\frac{2}{10}+\frac{7}{10}+\frac{5}{10}\right)=$

$26\frac{14}{10}=26\frac{7}{5}=27\frac{2}{5}$

79. $45-16\frac{2}{3}=44\frac{3}{3}-16\frac{2}{3}=28\frac{1}{3}$

81. $5\frac{31}{32}-1\frac{3}{16}=5\frac{31}{32}-1\frac{6}{32}=4\frac{25}{32}$

83. $8\frac{1}{2}+17\frac{7}{8}-6\frac{3}{4}=19+\left(\frac{1}{2}\cdot\frac{4}{4}+\frac{7}{8}-\frac{3}{4}\cdot\frac{2}{2}\right)=19+\left(\frac{4}{8}+\frac{7}{8}-\frac{6}{8}\right)=19\frac{5}{8}$

85. The LCM of 3, 5, 8, and 12 is 120.
    The LCM of 3, 5, 8, and 24 is 120.
    The LCM of 3, 5, 8, and 60 is 120.
    The LCM of 3, 5, 8, and 120 is 120.

87. The error is that 1 must be borrowed from the 16 is order to complete the subtraction. The correction is: $16 - 13\dfrac{1}{4} = 15 + \dfrac{4}{4} - 13\dfrac{1}{4} = 2\dfrac{3}{4}$

89. $12\dfrac{11}{12} + 22 + 8\dfrac{5}{8} = 42 + \dfrac{11}{12}\cdot\dfrac{2}{2} + \dfrac{5}{8}\cdot\dfrac{3}{3} = 42 + \dfrac{22}{24} + \dfrac{15}{24} = 42\dfrac{37}{24} = 43\dfrac{13}{24}$

91. $15\dfrac{3}{8} + 22\dfrac{1}{2} + 19\dfrac{5}{9} + 36\dfrac{2}{3} = 92 + \left(\dfrac{3}{8}\cdot\dfrac{9}{9} + \dfrac{1}{2}\cdot\dfrac{36}{36} + \dfrac{5}{9}\cdot\dfrac{8}{8} + \dfrac{2}{3}\cdot\dfrac{24}{24}\right) =$

$92 + \left(\dfrac{27}{72} + \dfrac{36}{72} + \dfrac{40}{72} + \dfrac{48}{72}\right) = 92\dfrac{151}{72} = 94\dfrac{7}{72}$

93. $6\dfrac{23}{25} - 5\dfrac{14}{15} = 1 + \left(\dfrac{23}{25}\cdot\dfrac{3}{3} - \dfrac{14}{15}\cdot\dfrac{5}{5}\right) = \dfrac{75}{75} + \left(\dfrac{69}{75} - \dfrac{70}{75}\right) = \dfrac{74}{75}$

95. $-\dfrac{121}{144}x^2 - \left(-\dfrac{13}{36}x^2\right) = -\dfrac{121}{144}x^2 + \dfrac{13}{36}\cdot\dfrac{4}{4}x^2 = -\dfrac{121}{144}x^2 + \dfrac{52}{144}x^2 = -\dfrac{69}{144}x^2 = -\dfrac{23}{48}x^2$

97. $\left(\dfrac{2}{3}x - \dfrac{3}{4}y\right) + \left(\dfrac{4}{3}x - \dfrac{5}{4}y\right) = \dfrac{6}{3}x - \dfrac{8}{4}y = 2x - 2y$

99. Add:

$\left(\dfrac{5}{3}x + \dfrac{3}{8}y - \dfrac{1}{2}\right) + \left(\dfrac{4}{5} - \dfrac{5}{6}x\right) = \left(\dfrac{5}{3}\cdot\dfrac{2}{2}x + \dfrac{3}{8}y - \dfrac{1}{2}\cdot\dfrac{5}{5}\right) + \left(\dfrac{4}{5}\cdot\dfrac{2}{2} - \dfrac{5}{6}x\right) =$

$\left(\dfrac{10}{6}x + \dfrac{3}{8}y - \dfrac{5}{10}\right) + \left(\dfrac{8}{10} - \dfrac{5}{6}x\right) = \dfrac{5}{6}x + \dfrac{3}{8}y + \dfrac{3}{10}$

101. $\dfrac{5}{8} + \dfrac{7}{8} + \dfrac{3}{8} + \dfrac{5}{8} + \dfrac{7}{8} = \dfrac{27}{8} = 3\dfrac{3}{8}$ points rise

103. $\dfrac{5}{4}\,\text{in} - \dfrac{1}{8}\,\text{in} = \dfrac{5}{4}\cdot\dfrac{2}{2}\,\text{in} - \dfrac{1}{8}\,\text{in} = \dfrac{10}{8}\,\text{in} - \dfrac{1}{8}\,\text{in} = \dfrac{9}{8}\,\text{in} = 1\dfrac{1}{8}\,\text{in}$

105. $\left(\dfrac{7}{8} + \dfrac{1}{16} + \dfrac{1}{2} + \dfrac{1}{8} + \dfrac{1}{4}\right)\text{in} = \left(\dfrac{7}{8}\cdot\dfrac{2}{2} + \dfrac{1}{16} + \dfrac{1}{2}\cdot\dfrac{8}{8} + \dfrac{1}{8}\cdot\dfrac{2}{2} + \dfrac{1}{4}\cdot\dfrac{4}{4}\right)\text{in} =$

$\left(\dfrac{14}{16} + \dfrac{1}{16} + \dfrac{8}{16} + \dfrac{2}{16} + \dfrac{4}{16}\right)\text{in} = \dfrac{29}{16}\,\text{in} = 1\dfrac{13}{16}\,\text{in}$

107. a. Westport − Freeport = $30\frac{3}{4} - 27\frac{1}{8} = 3 + \left(\frac{3}{4} \cdot \frac{2}{2} - \frac{1}{8}\right) = 3 + \left(\frac{6}{8} - \frac{1}{8}\right) = 3\frac{5}{8}$ in

   b. Salem + Forest Hills + Westview =

$$42\frac{1}{3} + 14\frac{5}{8} + 35\frac{1}{4} = 91 + \left(\frac{1}{3} \cdot \frac{8}{8} + \frac{5}{8} \cdot \frac{3}{3} + \frac{1}{4} \cdot \frac{6}{6}\right) =$$

$$91 + \left(\frac{8}{24} + \frac{15}{24} + \frac{6}{24}\right) = 91\frac{29}{24} = 92\frac{5}{24} \text{ in}$$

   c. 10(Salem − Forest Hills) =

$$10\left(42\frac{1}{3} - 14\frac{5}{8}\right) = 10\left(28 + \left(\frac{1}{3} - \frac{5}{8}\right)\right) = 10\left(28 + \left(\frac{1}{3} \cdot \frac{8}{8} - \frac{5}{8} \cdot \frac{3}{3}\right)\right) =$$

$$10\left(28 + \left(\frac{8}{24} - \frac{15}{24}\right)\right) = 10\left(28 + \left(-\frac{7}{24}\right)\right) = 10\left(27\frac{24}{24} + \left(-\frac{7}{24}\right)\right) =$$

$$10\left(27\frac{17}{24}\right) = 270\frac{170}{24} = 277\frac{2}{24} = 277\frac{1}{12} \text{ in}$$

   d. 2(Westview) − Salem =

$$2\left(35\frac{1}{4}\right) - 42\frac{1}{3} = 70\frac{1}{2} - 42\frac{1}{3} = 28 + \left(\frac{1}{2} \cdot \frac{3}{3} - \frac{1}{3} \cdot \frac{2}{2}\right) = 28 + \left(\frac{3}{6} - \frac{2}{6}\right) = 28\frac{1}{6} \text{ in}$$

109. $\left(\frac{1}{8} + \frac{1}{2} + \frac{1}{8}\right)$ in $= \left(\frac{1}{8} + \frac{1}{2} \cdot \frac{4}{4} + \frac{1}{8}\right)$ in $= \left(\frac{1}{8} + \frac{4}{8} + \frac{1}{8}\right)$ in $= \frac{6}{8}$ in $= \frac{3}{4}$ in

111. $\frac{5}{8}$ is a solution

$$x + \frac{2}{3} = \frac{31}{24}$$

$$\frac{5}{8} \cdot \frac{3}{3} + \frac{2}{3} \cdot \frac{8}{8} = \frac{31}{24}$$

$$\frac{15}{24} + \frac{16}{28} = \frac{31}{24}$$

$$\frac{31}{24} = \frac{31}{24}$$

113. The perimeter in feet is 2(10 ft + 6 ft) = 2(16 ft) = 32 ft

The perimeter in inches is $32\text{ ft}\left(\frac{12\text{ in}}{\text{ft}}\right) = 384\text{ in}$

The length of one brick and one seam of mortar is 8 in. $+ \frac{3}{8}$ in. $= 8\frac{3}{8}$ in.

$384\text{ in} \div 8\frac{3}{8}\text{ in} = \frac{384}{1}\text{ in} \cdot \frac{8}{67\text{ in}} = 45.8$ or 46 bricks

117

115. The probability that a card will be a heart or a club is: $\dfrac{13}{52}+\dfrac{13}{52}=\dfrac{26}{52}=\dfrac{1}{2}$

117. The probability of drawing any card less than a ten is the same as the probability of not drawing a ten or a Jack or a Queen or a King: $\quad 1-\dfrac{16}{52}=\dfrac{52}{52}-\dfrac{16}{52}=\dfrac{36}{52}=\dfrac{9}{13}$

119. $\dfrac{-4}{5}=-\dfrac{4}{5}$ because a -4 divided by $+5$ gives a negative result.

$\dfrac{4}{-5}=-\dfrac{4}{5}$ because a $+4$ divided by -5 gives a negative result.

121. $4\dfrac{2}{5}=4+\dfrac{2}{5}=4\cdot\dfrac{5}{5}+\dfrac{2}{5}=\dfrac{20}{5}+\dfrac{2}{5}=\dfrac{22}{5}$

123. The first five common multiples of 6 and 15 are: 30, 60, 90, 120, 150.
This list is also the multiples of 30. Note that 30 is the LCM for 6 and 15.

125. $2\dfrac{3}{8}-3\dfrac{3}{4}=\dfrac{19}{8}-\dfrac{15}{4}=\dfrac{19}{8}-\dfrac{15}{4}\cdot\dfrac{2}{2}=\dfrac{19}{8}-\dfrac{30}{8}=-\dfrac{11}{8}=-1\dfrac{3}{8}$

127. answers vary

129. -14 is a solution.

$$x^2-3x=238$$
$$(-14)^2-3(-14)=238$$
$$196+42=238$$
$$238=238$$

131. $\dfrac{374}{561}=\dfrac{2\cdot11\cdot17}{3\cdot11\cdot17}=\dfrac{2}{3}$

Section 5.6
Evaluating Expressions and Averages (Fractions)

1. Evaluate:
$$x + y$$
$$\frac{2}{3} + \frac{1}{6}$$
$$\frac{2}{3} \cdot \frac{2}{2} + \frac{1}{6}$$
$$\frac{4}{6} + \frac{1}{6}$$
$$\frac{5}{6}$$

3. Evaluate:
$$-x + y$$
$$-\frac{2}{3} + \frac{1}{6}$$
$$-\frac{2}{3} \cdot \frac{2}{2} + \frac{1}{6}$$
$$-\frac{4}{6} + \frac{1}{6}$$
$$-\frac{3}{6} = -\frac{1}{2}$$

5. Evaluate:

$$\frac{y}{x} = y \div x$$
$$\frac{1}{6} \div \frac{2}{3}$$
$$\frac{1}{6} \cdot \frac{3}{2}$$
$$\frac{1}{4}$$

7. Evaluate:
$$x^2 + y^2$$
$$\left(\frac{2}{3}\right)^2 + \left(\frac{1}{6}\right)^2$$
$$\frac{4}{9} + \frac{1}{36}$$
$$\frac{4}{9} \cdot \frac{4}{4} + \frac{1}{36}$$
$$\frac{16}{36} + \frac{1}{36}$$
$$\frac{17}{36}$$

9. Evaluate:

$$x - y + z$$

$$\frac{5}{12} - \frac{3}{16} + \frac{3}{20}$$

$$\frac{5}{12} \cdot \frac{20}{20} - \frac{3}{16} \cdot \frac{15}{15} + \frac{3}{20} \cdot \frac{12}{12}$$

$$\frac{100}{240} - \frac{45}{240} + \frac{36}{240}$$

$$\frac{91}{240}$$

11. Evaluate:

$$(x + y) \div z$$

$$\left(\frac{5}{12} + \frac{3}{16}\right) \div \frac{3}{20}$$

$$\left(\frac{5}{12} \cdot \frac{4}{4} + \frac{3}{16} \cdot \frac{3}{3}\right) \div \frac{3}{20}$$

$$\left(\frac{20}{48} + \frac{9}{48}\right) \div \frac{3}{20}$$

$$\frac{29}{48} \cdot \frac{20}{3}$$

$$\frac{145}{36}$$

13. Evaluate:

$$a + b + c$$

$$-\frac{4}{5} + \frac{7}{15} - \frac{5}{9}$$

$$-\frac{4}{5} \cdot \frac{9}{9} + \frac{7}{15} \cdot \frac{3}{3} - \frac{5}{9} \cdot \frac{5}{5}$$

$$-\frac{36}{45} + \frac{21}{45} - \frac{25}{45}$$

$$-\frac{40}{45} = -\frac{8}{9}$$

120

15. Evaluate:

$$(b-c) \div (a+c)$$

$$\left(\frac{7}{15} - \left(-\frac{5}{9}\right)\right) \div \left(-\frac{4}{5} + \left(-\frac{5}{9}\right)\right)$$

$$\left(\frac{7}{15} + \frac{5}{9}\right) \div \left(-\frac{4}{5} - \frac{5}{9}\right)$$

$$\left(\frac{7}{15} \cdot \frac{3}{3} + \frac{5}{9} \cdot \frac{5}{5}\right) \div \left(-\frac{4}{5} \cdot \frac{9}{9} - \frac{5}{9} \cdot \frac{5}{5}\right)$$

$$\left(\frac{21}{45} + \frac{25}{45}\right) \div \left(-\frac{36}{45} - \frac{25}{45}\right)$$

$$\frac{46}{45} \div \left(-\frac{61}{45}\right)$$

$$\frac{46}{45} \cdot \left(-\frac{45}{61}\right)$$

$$-\frac{46}{61}$$

17. Evaluate:

$$b^2 - a$$

$$\left(\frac{7}{15}\right)^2 - \left(-\frac{4}{5}\right)$$

$$\frac{49}{225} + \frac{4}{5}$$

$$\frac{49}{225} + \frac{4}{5} \cdot \frac{45}{45}$$

$$\frac{49}{225} + \frac{180}{225}$$

$$\frac{229}{225}$$

19. The average is:

$$\left(\frac{1}{6} + \frac{7}{12}\right) \div 2$$

$$\left(\frac{1}{6} \cdot \frac{2}{2} + \frac{7}{12}\right) \cdot \frac{1}{2}$$

$$\left(\frac{2}{12} + \frac{7}{12}\right) \cdot \frac{1}{2}$$

$$\left(\frac{9}{12}\right) \cdot \frac{1}{2}$$

$$\frac{3}{8}$$

21. The average is:

$$\left(\frac{1}{2}+\frac{1}{4}+\frac{3}{4}\right)\div 3$$

$$\left(\frac{1}{2}\cdot\frac{2}{2}+\frac{1}{4}+\frac{3}{4}\right)\cdot\frac{1}{3}$$

$$\left(\frac{2}{4}+\frac{1}{4}+\frac{3}{4}\right)\cdot\frac{1}{3}$$

$$\left(\frac{6}{4}\right)\cdot\frac{1}{3}$$

$$\frac{1}{2}$$

23. The average is:

$$\left(\frac{1}{2}+\frac{3}{4}+\frac{5}{8}+\frac{13}{16}\right)\div 4$$

$$\left(\frac{1}{2}\cdot\frac{8}{8}+\frac{3}{4}\cdot\frac{4}{4}+\frac{5}{8}\cdot\frac{2}{2}+\frac{13}{16}\right)\div 4$$

$$\left(\frac{8}{16}+\frac{12}{16}+\frac{10}{16}+\frac{13}{16}\right)\cdot\frac{1}{4}$$

$$\left(\frac{43}{16}\right)\cdot\frac{1}{4}$$

$$\frac{43}{64}$$

25. The average is:

$$\left(3\frac{1}{3}+4\frac{1}{6}+2\frac{2}{9}\right)\div 3$$

$$\left(\frac{10}{3}+\frac{25}{6}+\frac{20}{9}\right)\div 3$$

$$\left(\frac{10}{3}\cdot\frac{6}{6}+\frac{25}{6}\cdot\frac{3}{3}+\frac{20}{9}\cdot\frac{2}{2}\right)\div 3$$

$$\left(\frac{60}{18}+\frac{75}{18}+\frac{40}{18}\right)\cdot\frac{1}{3}$$

$$\frac{175}{18}\cdot\frac{1}{3}$$

$$\frac{175}{54}$$

27. $C = 2\pi r = 2\cdot\dfrac{22}{7}\cdot 7\,\text{cm} = 44\,\text{cm}$

29. $C = 2\pi r = 2\cdot\dfrac{22}{7}\cdot 14\,\text{ft} = 88\,\text{ft}$

31. $C = \pi d = \dfrac{22}{7}\cdot 5\,\text{m} = \dfrac{110}{7}\,\text{m}$

33. $V = \pi r^2 h = \dfrac{22}{7}\cdot(2\,\text{in})^2\cdot 7\,\text{in} = \dfrac{22}{7}\cdot 4\,\text{in}^2\cdot 7\,\text{in} = 88\,\text{in}^3$

35. $P = \pi r + 2r = \dfrac{22}{7} \cdot 10\,\text{ft} + 2 \cdot 10\,\text{ft} = \dfrac{220}{7}\,\text{ft} + \dfrac{7}{7} \cdot 20\,\text{ft} = \dfrac{220}{7}\,\text{ft} + \dfrac{140}{7}\,\text{ft} = \dfrac{360}{7}\,\text{ft}$

37. $A = \pi r^2 = \dfrac{22}{7} \cdot (9\,\text{in})^2 = \dfrac{22}{7} \cdot 81\,\text{in}^2 = \dfrac{1782}{7}\,\text{in}^2 = 254\dfrac{4}{7}\,\text{in}^2$

39.
$A = \dfrac{1}{2}bh + \dfrac{1}{2}\pi r^2 = \dfrac{1}{2}(9\,\text{in})(20\,\text{in}) + \dfrac{1}{2} \cdot \dfrac{22}{7} \cdot \left(\dfrac{9}{2}\,\text{in}\right)^2 =$

$90\,\text{in}^2 + \dfrac{891}{28}\,\text{in}^2 = 90\,\text{in}^2 + 31\dfrac{23}{28}\,\text{in}^2 = 121\dfrac{23}{28}\,\text{in}^2$

41. $V = \pi r^2 h = \dfrac{22}{7} \cdot \left(\dfrac{13}{2}\,\text{in}\right)^2 \left(4\,\text{ft} \cdot \dfrac{12\,\text{in}}{\text{ft}}\right) = \dfrac{22}{7} \cdot \dfrac{169}{4}\,\text{in}^2 (48\,\text{in}) = \dfrac{44{,}616\,\text{in}^3}{7} = 6373\dfrac{5}{7}\,\text{in}^3$

43. $A = \pi r^2 = \dfrac{22}{7} \cdot \left(\dfrac{7}{11}\,\text{in}\right)^2 = \dfrac{22}{7} \cdot \dfrac{49}{121}\,\text{in}^2 = \dfrac{14}{11}\,\text{in}^2 = 1\dfrac{3}{11}\,\text{in}^2$

45. $i = prt = \$50 \cdot \dfrac{6}{100} \cdot \dfrac{3}{4} = \$\dfrac{9}{4} = \$2.25$

47. The average is: $\left(\dfrac{11}{15} + \dfrac{11}{15} + \dfrac{11}{15} + \dfrac{8}{15} + \dfrac{8}{15}\right) \div 5 = \dfrac{49}{15} \cdot \dfrac{1}{5} = \dfrac{49}{75}$

49. The class average is

$\left(\dfrac{20}{20} + \dfrac{20}{20} + \dfrac{19}{20} + \dfrac{17}{20} + \dfrac{17}{20} + \dfrac{17}{20} + \dfrac{16}{20} + \dfrac{16}{20} + \dfrac{15}{20} + \dfrac{15}{20}\right) \div 10 = \dfrac{172}{20} \cdot \dfrac{1}{10} = \dfrac{43}{50}$

51. The average length of the salmon is:

$\left(23\dfrac{1}{4} + 31\dfrac{5}{8} + 42\dfrac{3}{4} + 28\dfrac{5}{8} + 35\dfrac{3}{4} + 40\right)\text{in} \div 6 =$

$\left(23\dfrac{2}{8} + 31\dfrac{5}{8} + 42\dfrac{6}{8} + 28\dfrac{5}{8} + 35\dfrac{6}{8} + 40\right)\text{in} \cdot \dfrac{1}{6} =$

$\left(199\dfrac{24}{8}\,\text{in}\right) \cdot \dfrac{1}{6} = (202\,\text{in}) \cdot \dfrac{1}{6} = 33\dfrac{2}{3}\,\text{in}$

53. The number of ounces of seafood in the carton is:

$$3\left(\frac{7}{2}\,\text{oz}\right)+5\left(\frac{27}{4}\,\text{oz}\right)+4\left(\frac{11}{2}\,\text{oz}\right)+4\left(\frac{21}{2}\,\text{oz}\right)=\left(\frac{21}{2}+\frac{135}{4}+\frac{22}{1}+\frac{42}{1}\right)\text{oz}=$$

$$\left(\frac{42}{4}+\frac{135}{4}+\frac{22}{1}+\frac{42}{1}\right)\text{oz}=\left(\frac{177}{4}+64\right)\text{oz}=\left(44\frac{1}{4}+64\right)\text{oz}=108\frac{1}{4}\,\text{oz}$$

The average cost per ounce is:

$$\$52\div108\frac{1}{4}\,\text{oz}=\$52\div\frac{433}{4}\,\text{oz}=\$52\cdot\frac{4}{433}\,\text{oz}=\frac{\$208}{433}\,\text{oz}=\$0.48$$

55. The volume of the cone is:

$$V=\frac{1}{3}\pi r^2 h=\frac{1}{3}\cdot\frac{22}{7}\cdot(17\,\text{mm})^2(36\,\text{mm})=\frac{1}{3}\cdot\frac{22}{7}\cdot289\,\text{mm}^2(36\,\text{mm})=$$

$$\frac{76{,}296}{7}\,\text{mm}^3=10{,}899\frac{3}{7}\,\text{mm}^3$$

57. The volume of a cone (with a diameter of 5 inches and a height of 8 inches) that is topped with a hemisphere of the same diameter is:

$$V=\frac{1}{3}\pi r^2 h+\frac{1}{2}\cdot\frac{4}{3}\pi r^3$$

$$V=\frac{1}{3}\cdot\frac{22}{7}\left(\frac{5}{2}\right)^2\left(\frac{8}{1}\right)+\frac{1}{2}\cdot\frac{4}{3}\cdot\frac{22}{7}\left(\frac{5}{2}\right)^3$$

$$V=\frac{1}{3}\cdot\frac{22}{7}\cdot\frac{25}{4}\left(\frac{8}{1}\right)+\frac{1}{2}\cdot\frac{4}{3}\cdot\frac{22}{7}\cdot\frac{125}{8}$$

$$V=\frac{1100}{21}+\frac{1375}{14}$$

$$V=\frac{1100}{21}\cdot\frac{2}{2}+\frac{1375}{42}$$

$$V=\frac{2200}{42}+\frac{1375}{42}$$

$$V=\frac{3575}{42}=85\frac{5}{42}\,\text{in}^3$$

59. The probability of drawing a jack of spades is: $\dfrac{4}{52}\cdot\dfrac{13}{52}=\dfrac{1}{52}$

61. Evaluate: $\dfrac{a-b}{5+c}=\dfrac{4\frac{1}{2}-\frac{1}{8}}{5+\frac{1}{4}}=\dfrac{\frac{9}{2}-\frac{1}{8}}{5+\frac{1}{4}}\cdot\dfrac{8}{8}=\dfrac{36-1}{40+2}=\dfrac{35}{42}=\dfrac{5}{6}$

63. answers vary

124

65. Evaluate:

$$x^3 + y^2 + z$$

$$\left(1\frac{1}{2}\right)^3 + \left(2\frac{2}{3}\right)^2 + 3\frac{3}{4}$$

$$\left(\frac{3}{2}\right)^3 + \left(\frac{8}{3}\right)^2 + \frac{15}{4}$$

$$\frac{27}{8} + \frac{64}{9} + \frac{15}{4}$$

$$\frac{27}{8} \cdot \frac{9}{9} + \frac{64}{9} \cdot \frac{8}{8} + \frac{15}{4} \cdot \frac{18}{18}$$

$$\frac{243}{72} + \frac{512}{72} + \frac{270}{72}$$

$$\frac{1025}{72} = 14\frac{17}{72}$$

67. The total amount paid by the three people to Acme is:

$$\$4500\left[3\frac{2}{5} + 1\frac{1}{2}\left(3\frac{2}{5}\right) + \frac{7}{8}\left(1\frac{1}{2}\right)\left(3\frac{2}{5}\right)\right]$$

$$\$4500\left[\frac{17}{5} + \frac{3}{2}\left(\frac{17}{5}\right) + \frac{7}{8}\left(\frac{3}{2}\right)\left(\frac{17}{5}\right)\right]$$

$$\$4500\left[\frac{17}{5} + \frac{51}{10} + \frac{357}{80}\right]$$

$$\$4500\left[\frac{17}{5} \cdot \frac{16}{16} + \frac{51}{10} \cdot \frac{8}{8} + \frac{357}{80}\right]$$

$$\$4500\left[\frac{272}{80} + \frac{408}{80} + \frac{357}{80}\right]$$

$$\$4500\left[\frac{1037}{80}\right]$$

$$\$58,331.25 \approx \$58,331$$

69. The prime factorization of $299 = 13 \cdot 23$

71. Solve:
$$2x - 8 = 10 - x$$
$$3x = 18$$
$$x = 6$$

73. Solve:
$$-2x + 12 = 3x + 32$$
$$-5x = 20$$
$$x = -4$$

Section 5.7
Solving Equations Involving Rational Numbers (Fractions)

1. Solve:

$$2x - \frac{1}{5} = \frac{1}{5}$$

$$5 \cdot 2x - \frac{1}{5} \cdot 5 = \frac{1}{5} \cdot 5$$

$$10x - 1 = 1$$

$$10x = 2$$

$$x = \frac{2}{10} = \frac{1}{5}$$

3. Solve:

$$\frac{1}{2}x + \frac{2}{3} = \frac{5}{3}$$

$$6 \cdot \frac{1}{2}x + \frac{2}{3} \cdot 6 = \frac{5}{3} \cdot 6$$

$$3x + 4 = 10$$

$$3x = 6$$

$$x = 2$$

5. Solve:

$$\frac{2}{3}x + 1 = \frac{5}{3}$$

$$3 \cdot \frac{2}{3}x + 1 \cdot 3 = \frac{5}{3} \cdot 3$$

$$2x + 3 = 5$$

$$2x = 2$$

$$x = 1$$

7. Solve:

$$\frac{1}{5}a - \frac{1}{3} = \frac{1}{5}$$

$$15 \cdot \frac{1}{5}a - \frac{1}{3} \cdot 15 = \frac{1}{5} \cdot 15$$

$$3a - 5 = 3$$

$$3a = 8$$

$$a = \frac{8}{3}$$

9. Solve:

$$\frac{1}{6} + \frac{1}{2}y = \frac{2}{3}$$

$$6 \cdot \frac{1}{6} + \frac{1}{2}y \cdot 6 = \frac{2}{3} \cdot 6$$

$$1 + 3y = 4$$

$$3y = 3$$

$$y = 1$$

11. Solve:

$$\frac{3}{7} + \frac{1}{4}c = \frac{1}{21}$$

$$84 \cdot \frac{3}{7} + \frac{1}{4}c \cdot 84 = \frac{1}{21} \cdot 84$$

$$36 + 21c = 4$$

$$21c = -32$$

$$c = -\frac{32}{21}$$

13. Solve:

$$\frac{1}{4}y - \frac{3}{2} = \frac{1}{8}$$

$$8 \cdot \frac{1}{4}y - \frac{3}{2} \cdot 8 = \frac{1}{8} \cdot 8$$

$$2y - 12 = 1$$

$$2y = 13$$

$$y = \frac{13}{2}$$

15. Solve:

$$\frac{5}{7} - \frac{1}{2}x = \frac{3}{14}$$

$$14 \cdot \frac{5}{7} - \frac{1}{2}x \cdot 14 = \frac{3}{14} \cdot 14$$

$$10 - 7x = 3$$

$$-7x = -7$$

$$x = 1$$

17. Solve:

$$\frac{3}{10}y - \frac{4}{5} = \frac{7}{20}$$

$$20 \cdot \frac{3}{10}y - \frac{4}{5} \cdot 20 = \frac{7}{20} \cdot 20$$

$$6y - 16 = 7$$

$$6y = 23$$

$$y = \frac{23}{6}$$

19. Solve:

$$\frac{a}{6} - \frac{a}{5} = \frac{2}{3} + \frac{1}{2}$$

$$\frac{a}{6} \cdot 30 - \frac{a}{5} \cdot 30 = \frac{2}{3} \cdot 30 + \frac{1}{2} \cdot 30$$

$$5a - 6a = 20 + 15$$

$$-a = 35$$

$$a = -35$$

21. Solve:

$$\frac{a}{9} - \frac{1}{2} = \frac{5}{18} + \frac{2}{3}$$

$$\frac{a}{9} \cdot 18 - \frac{1}{2} \cdot 18 = \frac{5}{18} \cdot 18 + \frac{2}{3} \cdot 18$$

$$2a - 9 = 5 + 12$$

$$2a - 9 = 17$$

$$2a = 26$$

$$a = 13$$

23. Solve:

$$\frac{3x}{10} + \frac{5x}{12} = \frac{5}{6} - \frac{1}{30}$$

$$\frac{3x}{10} \cdot 60 + \frac{5x}{12} \cdot 60 = \frac{5}{6} \cdot 60 - \frac{1}{30} \cdot 60$$

$$18x + 25x = 50 - 2$$

$$43x = 48$$

$$x = \frac{48}{43}$$

25. Solve:

$$\frac{13x}{20} - \frac{x}{3} = \frac{2}{5} - \frac{22}{30}$$

$$\frac{13x}{20} \cdot 60 - \frac{x}{3} \cdot 60 = \frac{2}{5} \cdot 60 - \frac{22}{30} \cdot 60$$

$$39x - 20x = 24 - 44$$

$$19x = -20$$

$$x = -\frac{20}{19}$$

27. Solve:

$$\frac{6c}{7} - 12 = \frac{3}{5} + 5$$

$$\frac{6c}{7} \cdot 35 - 12 \cdot 35 = \frac{3}{5} \cdot 35 + 5 \cdot 35$$

$$30c - 420 = 21 + 175$$

$$30c - 420 = 196$$

$$30c = 616$$

$$c = \frac{616}{30} = \frac{308}{15}$$

29. Solve:

$$-\frac{11b}{12} - \frac{2}{3} = -\frac{5}{6} + 3$$

$$-\frac{11b}{12} \cdot 12 - \frac{2}{3} \cdot 12 = -\frac{5}{6} \cdot 12 + 3 \cdot 12$$

$$-11b - 8 = -10 + 36$$

$$-11b - 8 = 26$$

$$-11b = 34$$

$$b = -\frac{34}{11}$$

31. Solve:

$$\frac{2a}{18} + \frac{7a}{72} = \frac{1}{18} + \frac{7}{9}$$

$$\frac{2a}{18} \cdot 72 + \frac{7a}{72} \cdot 72 = \frac{1}{18} \cdot 72 + \frac{7}{9} \cdot 72$$

$$8a + 7a = 4 + 56$$

$$15a = 60$$

$$a = 4$$

33. Solve:

$$\frac{9}{8} - \frac{7z}{20} = \frac{3}{50} + \frac{13}{24}$$

$$\frac{9}{8} \cdot 600 - \frac{7z}{20} \cdot 600 = \frac{3}{50} \cdot 600 + \frac{13}{24} \cdot 600$$

$$675 - 210z = 36 + 325$$

$$675 - 210z = 361$$

$$-210z = -314$$

$$z = \frac{-314}{-210} = \frac{157}{105}$$

35. Solve:

$$\frac{3}{8} + \frac{5a}{6} - \frac{7a}{24} + \frac{5}{9} = 0$$

$$\frac{3}{8} \cdot 72 + \frac{5a}{6} \cdot 72 - \frac{7a}{24} \cdot 72 + \frac{5}{9} \cdot 72 = 0 \cdot 72$$

$$27 + 60a - 21a + 40 = 0$$

$$39a + 67 = 0$$

$$39a = -67$$

$$a = -\frac{67}{39}$$

37. Solve:

$$\frac{1}{5}\left(\frac{8}{3}+\frac{4y}{15}\right)=\frac{11}{25}$$

$$\frac{8}{15}+\frac{4y}{75}=\frac{11}{25}$$

$$\frac{8}{15}\cdot 75+\frac{4y}{75}\cdot 75=\frac{11}{25}\cdot 75$$

$$40+4y=33$$

$$4y=-7$$

$$y=-\frac{7}{4}$$

39. Solve:

$$\frac{2}{5}\left(\frac{2p}{3}-\frac{5}{8}\right)=-\frac{1}{20}$$

$$\frac{4p}{15}-\frac{1}{4}=-\frac{1}{20}$$

$$\frac{4p}{15}\cdot 60-\frac{1}{4}\cdot 60=-\frac{1}{20}\cdot 60$$

$$16p-15=-3$$

$$16p=12$$

$$p=\frac{12}{-16}=\frac{3}{4}$$

41. Solve:

$$3x-\frac{7}{8}=5x+\frac{2}{3}$$

$$3x\cdot 24-\frac{7}{8}\cdot 24=5x\cdot 24+\frac{2}{3}\cdot 24$$

$$72x-21=120x+16$$

$$-37=48x$$

$$-\frac{37}{48}=x$$

43. Solve:

$$\frac{23}{24}x-\frac{2}{3}x=\frac{11}{12}x+10$$

$$\frac{23}{24}\cdot 24x-\frac{2}{3}\cdot 24x=\frac{11}{12}\cdot 24x+10\cdot 24$$

$$23x-16x=22x+240$$

$$7x=22x+240$$

$$-15x=240$$

$$x=-16$$

45. Solve:

$$\frac{7}{18}c-\frac{5}{27}=-\frac{8}{9}-\frac{1}{9}c$$

$$54\cdot\frac{7}{18}c-\frac{5}{27}\cdot 54=-\frac{8}{9}\cdot 54-54\cdot\frac{1}{9}c$$

$$21c-10=-48-6c$$

$$27c=-38$$

$$c=-\frac{38}{27}$$

47. Solve:

$$\frac{1}{12}\left(\frac{17}{3}x+\frac{11}{4}\right)=\frac{1}{24}(13x+7)$$

$$24\cdot\frac{1}{12}\left(1\frac{7}{3}x+\frac{11}{4}\right)=24\cdot\frac{1}{24}(13x+7)$$

$$2\left(\frac{17}{3}x+\frac{11}{4}\right)=13x+7$$

$$\frac{34}{3}x+\frac{11}{2}=13x+7$$

$$6\cdot\frac{34}{3}x+6\cdot\frac{11}{2}=6\cdot 13x+6\cdot 7$$

$$68x+33=78x+42$$

$$-9=10x$$

$$-\frac{9}{10}=x$$

49. The width of the rectangle is:
$$P = 2w + 2l$$

$$6\frac{2}{3}\,\text{in} = 2w + 2\left(2\frac{1}{2}\,\text{in}\right)$$

$$\frac{20}{3}\,\text{in} = 2w + 2\left(\frac{5}{2}\,\text{in}\right)$$

$$\frac{20}{3}\,\text{in} = 2w + 5\,\text{in}$$

$$3 \cdot \frac{20}{3}\,\text{in} = 3 \cdot 2w + 3 \cdot 5\,\text{in}$$

$$20\,\text{in} = 6w + 15\,\text{in}$$

$$5\,\text{in} = 6w$$

$$\frac{5}{6}\,\text{in} = w$$

51. The length of the rectangle is:
$$P = 2w + 2l$$

$$36\,\text{in} = 2\left(\frac{l}{2}\right) + 2l$$

$$36\,\text{in} = l + 2l$$

$$36\,\text{in} = 3l$$

$$12\,\text{in} = l$$

53. The number is:
$$\frac{x}{3} + \frac{2}{5} = \frac{26}{15}$$

$$15 \cdot \frac{x}{3} + 15 \cdot \frac{2}{5} = 15 \cdot \frac{26}{15}$$

$$5x + 6 = 26$$

$$5x = 20$$

$$x = 4$$

55. The base of the triangle is:

$$A = \frac{1}{2}bh$$

$$10\frac{1}{16}\,\text{in}^2 = \frac{1}{2}b\left(5\frac{3}{4}\,\text{in}\right)$$

$$\frac{161}{16}\,\text{in}^2 = \frac{1}{2}b\left(\frac{23}{4}\,\text{in}\right)$$

$$\frac{161}{16}\,\text{in}^2 = b\left(\frac{23}{8}\,\text{in}\right)$$

$$16\cdot\frac{161}{16}\,\text{in}^2 = b\cdot16\left(\frac{23}{8}\,\text{in}\right)$$

$$161\,\text{in}^2 = b(46\,\text{in})$$

$$\frac{161\,\text{in}^2}{46\,\text{in}} = b$$

$$3\frac{1}{2}\,\text{in} = b$$

57. The height of the cylinder is:

$$V = \pi r^2 h$$

$$770\,\text{cm}^3 = \frac{22}{7}(5\,\text{cm})^2 h$$

$$770\,\text{cm}^3 = \frac{22}{7}(25\,\text{cm}^2)h$$

$$770\,\text{cm}^3 = (\frac{550}{7}\,\text{cm}^2)h$$

$$770\,\text{cm}(\frac{7}{550}) = (\frac{550}{7})(\frac{7}{550})h$$

$$\frac{5390}{550}\,\text{cm} = \frac{539}{55}\,\text{cm} = h$$

59. The probability is: $\dfrac{59}{100} - \dfrac{11}{20} = \dfrac{59}{100} - \dfrac{11}{20}\cdot\dfrac{5}{5} = \dfrac{59}{100} - \dfrac{55}{100} = \dfrac{4}{100} = \dfrac{1}{25}$

61. answers vary

63. The equation is $\dfrac{3}{4}x - \dfrac{5}{12} = -\dfrac{1}{6} + \dfrac{7}{32}$

$$\dfrac{3}{4}\left(\dfrac{5}{8}\right) - \dfrac{5}{12} = -\dfrac{1}{6} + N$$

$$\dfrac{15}{32} - \dfrac{5}{12} = -\dfrac{1}{6} + N$$

$$96 \cdot \dfrac{15}{32} - \dfrac{5}{12} \cdot 96 = -\dfrac{1}{6} \cdot 96 + 96N$$

$$45 - 40 = -16 + 96N$$

$$5 = -16 + 96N$$

$$\dfrac{21}{96} = N$$

$$\dfrac{7}{32} = N$$

65. group activity

67. $\dfrac{72xy^2}{81xy} = \dfrac{8y}{9}$

69. $\left(-\dfrac{25y}{32}\right)\left(\dfrac{8y}{15}\right) = -\dfrac{5y^2}{12}$

Chapter 5
True-False Concept Review

1. False – If the numerator and the denominator of a fraction are both positive *and* the fraction is improper *and* the fraction is not equivalent to 1, then the denominator is less than the numerator.

2. True

3. False – The proper fraction $\dfrac{5}{6}$ can be changed to the mixed number $0\dfrac{5}{6}$.

4. False – The integer 7 is a factor and a divisor of 35.

5. True

6. False – Every whole number greater than 1 is either prime or composite.

7. True

8. False – When a fraction is reduced, its value remains the same.

9. False – Mixed numbers may or may not contain fractional components that can be reduced.

10. True

11. False – The numerator of the sum of two fractions is the sum of the two numerators only when the fractions already have a common denominator.

12. True

13. True

14. True

15. False – To find the reciprocal of a mixed number, first change the mixed number to an improper fraction and then interchange the numerator and the denominator.

16. True

17. False – All rational numbers *can* be written in fraction form.

18. True

19. False – The opposite of a positive fraction will have a negative numerator and the opposite of a negative fraction will have a positive numerator.

20. True

21. True

22. False – Building fractions is useful for adding, subtracting and comparing fractions.

23. False – A fraction renamed by building has the same value as the original fraction.

24. True

25. False – The smallest divisor, other than 1, of the LCM of three numbers is the smallest prime number that is a factor of the one of the three numbers.

26. True

27. False – The LCM of a list of numbers is either equal to or larger than the largest number in the list.

28. False – The denominator of the sum of two fractions is the LCM of the denominators of the two fractions.

29. True

30. True

31. True

32. True

33. False – Mixed numbers are added by adding the whole number portions and the fraction portions (which may involve "carrying").

34. True

35. True

36. False – The order of operations is the same for any kind of number.

37. True

38. False – The multiplication property of equality can be used to eliminate the fractions in an equation.

39. True

40. False – All rational numbers can be written in fraction form.

41. False – Two fractions are equivalent if they represent the same number.

42. False – Of two fractions with the same denominators, the fraction with the larger numerator has the larger value.

43. False – The commutative and the associative properties are used to group the whole numbers and to group the fractions when adding mixed number.

44. False – To add two fractions with the same denominators, add only the numerators.

45. False – The sum of two positive fractions may be smaller or larger than either of the fractions, depending on whether they are positive or negative.

46. False – The product of the numbers is a common multiple, but it may not be the *least* common multiple.

47. True

48. True

Chapter 5
Review

1. $\dfrac{3}{8}$

3. $\dfrac{7}{8}$

5. $\dfrac{11}{7}$

7. The improper fractions are: $\dfrac{7}{7}, \dfrac{27}{27}, -\dfrac{149}{148}$

9. The improper fractions are: $\dfrac{8}{7}, -\dfrac{10}{9}, \dfrac{11}{10}, -\dfrac{12}{12}$

11. The opposite of $\dfrac{7}{9}$ is $-\dfrac{7}{9}$

13. The opposite of $\dfrac{-13}{7}$ is $\dfrac{13}{7}$

15. The opposite of $-\dfrac{8}{3}$ is $\dfrac{8}{3}$

17. $\dfrac{19}{4} = 4\dfrac{3}{4}$

19. $\dfrac{145}{7} = 20\dfrac{5}{7}$

21. $8\dfrac{2}{3} = \dfrac{26}{3}$

23. $101\dfrac{4}{7} = \dfrac{711}{7}$

25. $51\dfrac{4}{11} = \dfrac{565}{11}$

27. The first five multiples of 19 are: 19, 38, 57, 76, 95

29. The first five multiples of 33 are: 33, 66, 99, 132, 165

31. 112 is a multiple of 7 since $112 = 7 \cdot 16$

33. 156 is a multiple of 13 since $156 = 13 \cdot 12$

35. 1197 is a multiple of 19 since $1197 = 19 \cdot 63$

37. 5231 is not divisible by any of 2, 3, 5, or 10

39. 5233 is not divisible by any of 2, 3, 5, or 10

41. The factors of 124 are: 1, 2, 4, 31, 62, 124

43. The factors of 630 are: 1, 2, 3, 5, 6, 7, 9, 10, 14, 15, 18, 21, 30, 35, 42, 45, 63, 70, 90, 105, 126, 210, 315, 630

45. The factors of 384 are: 1, 2, 3, 4, 6, 8, 12, 16, 24, 32, 48, 64, 96, 128, 192, 384

47. 213 is a composite number ($213 = 3 \times 71$)

49. 181 is a prime number

51. The prime factorization of 192 is $2^6 \cdot 3$

53. The prime factorization of 610 is $2 \cdot 5 \cdot 61$

55. The prime factorization of 1224 is $2^3 \cdot 3^2 \cdot 17$

57. $\dfrac{72}{96} = \dfrac{3}{4}$

59. $-\dfrac{30ab}{105} = -\dfrac{2ab}{7}$

61. $\dfrac{6}{7} \cdot \dfrac{2}{2} = \dfrac{12}{14}$ $\qquad$ $\dfrac{6}{7} \cdot \dfrac{3}{3} = \dfrac{18}{21}$ $\qquad$ $\dfrac{6}{7} \cdot \dfrac{4}{4} = \dfrac{24}{28}$ $\qquad$ $\dfrac{6}{7} \cdot \dfrac{5}{5} = \dfrac{30}{35}$

63. $\dfrac{5}{9} \cdot \dfrac{8}{8} = \dfrac{40}{72}$

65. $\dfrac{4y}{7x} \cdot \dfrac{3z}{3z} = \dfrac{12yz}{21xz}$

67. $-\dfrac{36}{60} \cdot \dfrac{45}{63} = -\dfrac{3}{7}$

69. $\left(2\dfrac{1}{7}\right)\left(4\dfrac{3}{10}\right) = \left(\dfrac{15}{7}\right)\left(\dfrac{43}{10}\right) = \dfrac{129}{14} = 9\dfrac{3}{14}$

71. $\dfrac{16}{25} \div \dfrac{8}{15} = \dfrac{16}{25} \cdot \dfrac{15}{8} = \dfrac{6}{5} = 1\dfrac{1}{5}$

73. $\dfrac{24}{49}d \div \dfrac{6}{7} = \dfrac{24}{49}d \cdot \dfrac{7}{6} = \dfrac{4d}{7}$

75. $\dfrac{5}{4} \div \dfrac{5}{16} = \dfrac{5}{4} \cdot \dfrac{16}{5} = 4$ turns

77. $124 \text{ pt} \cdot \dfrac{\text{gal}}{8 \text{ pt}} = 15.5 \text{ gal}$

76.

79. $\dfrac{40 \text{ km}}{\text{hr}} \cdot \dfrac{1000 \text{ m}}{\text{km}} \cdot \dfrac{\text{hr}}{3600 \text{ sec}} = 11\dfrac{1}{9} \text{ m/sec}$

81. The LCM is $40 = 2^3 \cdot 5$
$$5 = 5$$
$$10 = 2 \cdot 5$$
$$40 = 2^3 \cdot 5$$

83. The LCM is $180 = 2^2 \cdot 3^2 \cdot 5$
$$15 = 3 \cdot 5$$
$$60 = 2^2 \cdot 3 \cdot 5$$
$$90 = 2 \cdot 3^2 \cdot 5$$

85. The LCM is $2280 = 2^3 \cdot 3 \cdot 5 \cdot 19$
$$57 = 3 \cdot 19$$
$$95 = 5 \cdot 19$$
$$152 = 2^3 \cdot 19$$
$$190 = 2 \cdot 5 \cdot 19$$

87. $\dfrac{8}{3} > \dfrac{9}{4}$
$$\dfrac{8}{3} \cdot \dfrac{4}{4} = \dfrac{32}{12}$$
$$\dfrac{9}{4} \cdot \dfrac{3}{3} = \dfrac{27}{12}$$

89. $\dfrac{15}{17} > \dfrac{29}{33}$
$$\dfrac{15}{17} \cdot \dfrac{33}{33} = \dfrac{495}{561}$$
$$\dfrac{29}{33} \cdot \dfrac{17}{17} = \dfrac{493}{561}$$

91. $\dfrac{5}{17} + \dfrac{11}{17} = \dfrac{16}{17}$

93. $\dfrac{33}{40} + \left(-\dfrac{9}{40}\right) = \dfrac{24}{40} = \dfrac{3}{5}$

95. $\dfrac{12}{25}a + \left(-\dfrac{7}{25}\right)a = \dfrac{5}{25}a = \dfrac{1}{5}a$

97. Add:

$$\frac{3}{4}+\frac{7}{18}$$

$$\frac{3}{4}\cdot\frac{9}{9}+\frac{7}{18}\cdot\frac{2}{2}$$

$$\frac{27}{36}+\frac{14}{36}$$

$$\frac{41}{36}=1\frac{5}{36}$$

99. Add:

$$\frac{17}{18}b+\frac{11}{45}b$$

$$\frac{17}{18}\cdot\frac{5}{5}b+\frac{11}{45}\cdot\frac{2}{2}b$$

$$\frac{85}{90}b+\frac{22}{90}b$$

$$\frac{107}{90}b$$

101. Subtract:

$$\frac{23}{25}-\frac{6}{50}$$

$$\frac{23}{25}\cdot\frac{2}{2}-\frac{6}{50}$$

$$\frac{46}{50}-\frac{6}{50}$$

$$\frac{40}{50}=\frac{4}{5}$$

103. Subtract:

$$-\frac{5}{9}-\left(-\frac{1}{18}\right)$$

$$-\frac{5}{9}\cdot\frac{2}{2}-\left(-\frac{1}{18}\right)$$

$$-\frac{10}{18}-\left(-\frac{1}{18}\right)$$

$$-\frac{10}{18}+\left(+\frac{1}{18}\right)$$

$$-\frac{9}{18}=-\frac{1}{2}$$

105. Subtract:

$$\frac{7}{20}y-\left(-\frac{1}{15}y\right)$$

$$\frac{7}{20}y\cdot\frac{3}{3}-\left(-\frac{1}{15}y\cdot\frac{4}{4}\right)$$

$$\frac{21}{60}y-\left(-\frac{4}{60}y\right)$$

$$\frac{21}{60}y+\left(+\frac{4}{60}y\right)$$

$$\frac{25}{60}y=\frac{5}{12}y$$

107. Add:

$$9\frac{8}{9}+12\frac{5}{6}$$

$$21+\frac{8}{9}\cdot\frac{2}{2}+\frac{5}{6}\cdot\frac{3}{3}$$

$$21+\frac{16}{18}+\frac{15}{18}$$

$$21+\frac{31}{18}$$

$$21+1+\frac{13}{18}$$

$$22\frac{13}{18}$$

109. Add:

$$6\frac{1}{8}+19\frac{7}{12}+42\frac{1}{15}$$

$$67+\frac{1}{8}\cdot\frac{15}{15}+\frac{7}{12}\cdot\frac{10}{10}+\frac{1}{15}\cdot\frac{8}{8}$$

$$67+\frac{15}{120}+\frac{70}{120}+\frac{8}{120}$$

$$67+\frac{93}{120}$$

$$67\frac{31}{40}$$

111. Subtract:

$$6\frac{3}{8}-3\frac{1}{8}$$

$$3\frac{2}{8}$$

$$3\frac{1}{4}$$

113. Subtract:

$$8\frac{3}{7}-5\frac{9}{14}$$

$$2+1+\left(\frac{3}{7}\cdot\frac{2}{2}-\frac{9}{14}\right)$$

$$2+\frac{14}{14}+\left(\frac{6}{14}-\frac{9}{14}\right)$$

$$2+\left(\frac{20}{14}-\frac{9}{14}\right)$$

$$2\frac{11}{14}$$

115. Subtract:

$$459\frac{17}{54}-342\frac{41}{135}$$

$$117+\left(\frac{17}{54}\cdot\frac{5}{5}-\frac{41}{135}\cdot\frac{2}{2}\right)$$

$$117+\left(\frac{85}{270}-\frac{82}{270}\right)$$

$$117\frac{3}{270}=117\frac{1}{90}$$

117. Evaluate:

$$\frac{3}{4}a-\frac{2}{5}b+ab$$

$$\frac{3}{4}\cdot\frac{8}{9}-\frac{2}{5}\cdot\frac{5}{4}+\frac{8}{9}\cdot\frac{5}{4}$$

$$\frac{2}{3}-\frac{1}{2}+\frac{10}{9}$$

$$\frac{2}{3}\cdot\frac{6}{6}-\frac{1}{2}\cdot\frac{9}{9}+\frac{10}{9}\cdot\frac{2}{2}$$

$$\frac{12}{18}-\frac{9}{18}+\frac{20}{18}$$

$$\frac{3}{18}+\frac{20}{18}$$

$$\frac{23}{18}=1\frac{5}{18}$$

119. Evaluate:

$$A = p + prt$$

$$A = \$6000 + \$6000(7\%)\left(1\frac{1}{2}\text{ yr}\right)$$

$$A = \$6000 + \$6000\left(\frac{7}{100}\right)\left(\frac{3}{2}\text{ yr}\right)$$

$$A = \$6000 + \$630$$

$$a = \$6630$$

121. The average is:

$$\frac{\frac{1}{4}+\frac{3}{8}+\frac{15}{16}}{3}$$

$$\frac{\frac{1}{4}+\frac{3}{8}+\frac{15}{16}}{3}\cdot\frac{16}{1}$$

$$\frac{4+6+15}{48}$$

$$\frac{25}{48}$$

123. The average is:

$$\frac{\frac{1}{2}+\frac{1}{4}+\frac{9}{8}+\frac{19}{16}}{4}$$

$$\frac{\frac{1}{2}+\frac{1}{4}+\frac{9}{8}+\frac{19}{16}}{4}\cdot\frac{16}{1}$$

$$\frac{8+4+18+19}{64}$$

$$\frac{49}{64}$$

125. The average is:

$$\frac{3\frac{1}{6}+5\frac{1}{2}+6\frac{3}{4}+8\frac{5}{6}}{4}$$

$$\frac{\frac{19}{6}+\frac{11}{2}+\frac{27}{4}+\frac{53}{6}}{4}\cdot\frac{12}{1}$$

$$\frac{38+66+81+106}{48}$$

$$\frac{291}{48}=\frac{97}{16}=6\frac{1}{16}$$

127. The area is: $A=\pi r^2=\frac{22}{7}(8\,\text{in})^2=\frac{22}{7}\cdot\frac{64}{1}\,\text{in}^2=\frac{1408}{7}\,\text{in}^2=201\frac{1}{7}\,\text{in}^2$

129. The volume is:

$$V=\frac{4}{3}\pi r^3=\frac{4}{3}\left(\frac{22}{7}\right)\left(\frac{25}{2}\,\text{m}\right)^3=\frac{4}{3}\left(\frac{22}{7}\right)\left(\frac{15625}{8}\,\text{m}^3\right)=\frac{171875}{21}\,\text{m}^3=8184\frac{11}{21}\,\text{m}^3$$

131. Solve:

$$\frac{3}{7}x-\frac{5}{14}=\frac{1}{2}$$

$$14\cdot\frac{3}{7}x-\frac{5}{14}\cdot14=\frac{1}{2}\cdot14$$

$$6x-5=7$$

$$6x=12$$

$$x=2$$

133. Solve:

$$\frac{a}{3} - \frac{11}{12} + \frac{5a}{6} = \frac{17}{12}$$

$$\frac{a}{3} \cdot 12 - \frac{11}{12} \cdot 12 + \frac{5a}{6} \cdot 12 = \frac{17}{12} \cdot 12$$

$$4a - 11 + 10a = 17$$

$$14a = 28$$

$$a = 2$$

135. The number is:

$$\frac{x}{21} - \frac{3}{7} = -\frac{4}{21}$$

$$\frac{x}{21} \cdot 21 - \frac{3}{7} \cdot 21 = -\frac{4}{21} \cdot 21$$

$$x - 9 = -4$$

$$x = 5$$

Chapter 5
Test

1. The average is:

$$\frac{\frac{1}{5} + \frac{2}{3} + \frac{5}{6} + \frac{7}{10}}{4} = \frac{\frac{1}{5} + \frac{2}{3} + \frac{5}{6} + \frac{7}{10}}{4} \cdot \frac{30}{30} = \frac{6 + 20 + 25 + 21}{120} = \frac{72}{120} = \frac{3}{5}$$

2. $\dfrac{81}{15} = 5\dfrac{6}{15} = 5\dfrac{3}{5}$

3. $5\dfrac{3}{8} = \dfrac{43}{8}$

4. The first five multiples of 13 are 13, 26, 39, 52, 65.

5. 72 is a multiple of 8 since $8(9) = 72$

6. The factors of 72 are 1, 2, 3, 4, 6, 8, 9, 12, 18, 36, 72.

7. 113 is a prime number.

8. The prime factorization of 260 is $2 \cdot 2 \cdot 5 \cdot 13 = 2^2 \cdot 5 \cdot 13$

9. $\dfrac{36b^2}{54b} = \dfrac{2b}{3}$

10. $\dfrac{3}{16}\left(-\dfrac{8}{9}\right)\left(\dfrac{15}{4}\right) = -\dfrac{5}{8}$

11. $\left(-\dfrac{8d}{45}\right)\left(-\dfrac{15d}{34}\right) = \dfrac{3d^2}{17}$

12. $\left(2\dfrac{3}{4}\right)\left(3\dfrac{3}{5}\right) = \left(\dfrac{11}{4}\right)\left(\dfrac{18}{5}\right) = \dfrac{99}{10} = 9\dfrac{9}{10}$

13. $\left(-\dfrac{30x}{25}\right) \div \left(-\dfrac{6x}{15}\right) = \left(-\dfrac{30x}{25}\right)\left(-\dfrac{15}{6x}\right) = 3$

14. $\left(3\dfrac{5}{9}\right) \div \left(1\dfrac{1}{3}\right) = \left(\dfrac{32}{9}\right) \div \left(\dfrac{4}{3}\right) = \left(\dfrac{32}{9}\right)\left(\dfrac{3}{4}\right) = \dfrac{8}{3} = 2\dfrac{2}{3}$

15. $\dfrac{5}{8} \cdot \dfrac{8}{8} = \dfrac{40}{64}$

16. The LCM is 216

$8 = 2^3$

$27 = 3^3$

$36 = 2^2 \cdot 3^2$

$\text{LCM} = 2^3 \cdot 3^3$

17. $\dfrac{4}{5} < \dfrac{8}{9}$ is true

$\dfrac{4}{5} \cdot \dfrac{9}{9} = \dfrac{36}{45}$

$\dfrac{8}{9} \cdot \dfrac{5}{5} = \dfrac{40}{45}$

18. $\dfrac{5}{12}a + \dfrac{1}{12}a = \dfrac{6}{12}a = \dfrac{1}{2}a$

19. Add:

$$\dfrac{4}{9} + \left(-\dfrac{5}{12}\right) + \dfrac{3}{4}$$

$$\dfrac{4}{9} \cdot \dfrac{4}{4} + \left(-\dfrac{5}{12} \cdot \dfrac{3}{3}\right) + \dfrac{3}{4} \cdot \dfrac{9}{9}$$

$$\dfrac{16}{36} + \left(-\dfrac{15}{36}\right) + \dfrac{27}{36}$$

$$\dfrac{28}{36} = \dfrac{7}{9}$$

20. Add:

$$\dfrac{4}{7}y + \dfrac{2}{21}y + \dfrac{3}{14}y$$

$$\dfrac{4}{7} \cdot \dfrac{6}{6}y + \dfrac{2}{21} \cdot \dfrac{2}{2}y + \dfrac{3}{14} \cdot \dfrac{3}{3}y$$

$$\dfrac{24}{42}y + \dfrac{4}{42}y + \dfrac{9}{42}y$$

$$\dfrac{37y}{42}$$

21. The difference is:

$$\frac{11}{15} - \frac{1}{3}$$

$$\frac{11}{15} - \frac{1}{3} \cdot \frac{5}{5}$$

$$\frac{11}{15} - \frac{5}{15}$$

$$\frac{6}{15} = \frac{2}{5}$$

22. Subtract:

$$\frac{2}{2} \cdot \frac{7}{18} ab - \left(-\frac{1}{4} \cdot \frac{9}{9} ab \right)$$

$$\frac{14}{36} ab - \left(-\frac{9}{36} ab \right)$$

$$\frac{14}{36} ab + \left(+\frac{9}{36} ab \right)$$

$$\frac{23}{36} ab$$

23. Add:

$$5\frac{5}{6} + 3\frac{3}{10}$$

$$8 + \frac{5}{6} \cdot \frac{5}{5} + \frac{3}{10} \cdot \frac{3}{3}$$

$$8 + \frac{25}{30} + \frac{9}{30}$$

$$8 \frac{34}{30}$$

$$8 + 1 + \frac{4}{30}$$

$$9 \frac{2}{15}$$

24. Subtract:

$$8\frac{3}{10} - 5\frac{8}{15}$$

$$3 + \frac{3}{10} \cdot \frac{3}{3} - \frac{8}{15} \cdot \frac{2}{2}$$

$$2 + \frac{30}{30} + \frac{9}{30} - \frac{16}{30}$$

$$2 + \frac{39}{30} - \frac{16}{30}$$

$$2\frac{23}{30}$$

25. Evaluate:

$$xy - xz + y$$

$$\left(\frac{3}{4} \right)\left(\frac{5}{8} \right) - \left(\frac{3}{4} \right)\left(\frac{1}{3} \right) + \frac{5}{8}$$

$$\frac{15}{32} - \frac{1}{4} + \frac{5}{8}$$

$$\frac{15}{32} - \frac{1}{4} \cdot \frac{8}{8} + \frac{5}{8} \cdot \frac{4}{4}$$

$$\frac{15}{32} - \frac{8}{32} + \frac{20}{32}$$

$$\frac{27}{32}$$

26. Solve:

$$\frac{2}{3} y - \frac{1}{6} = \frac{3}{4}$$

$$12 \cdot \frac{2}{3} y - \frac{1}{6} \cdot 12 = \frac{3}{4} \cdot 12$$

$$8y - 2 = 9$$

$$8y = 11$$

$$y = \frac{11}{8}$$

27. The area of the triangle is:

$$A = \frac{1}{2}bh = \frac{1}{2}hb$$

$$A = \frac{1}{2}\left(2\frac{3}{4}\text{ in}\right)\left(3\frac{3}{8}\text{ in}\right)$$

$$A = \frac{1}{2}\left(\frac{11}{4}\text{ in}\right)\left(\frac{27}{8}\text{ in}\right)$$

$$A = \frac{297}{64}\text{ in}^2$$

$$A = 4\frac{41}{64}\text{ in}^2$$

28. The time to make one pin is: $68\frac{1}{2}\text{ min} \div 4 = \frac{137}{2}\text{ min} \cdot \frac{1}{4} = \frac{137}{8}\text{ min} = 17\frac{1}{8}\text{ min}$

29. The number of truckloads of hay in the railroad car is:

$$120\frac{1}{2} \div 6\frac{3}{4} = \frac{241}{2} \div \frac{27}{4} = \frac{241}{2} \cdot \frac{4}{27} = \frac{482}{27} = 17\frac{23}{27}$$

30. The amount that Jill will have to cut off the board is:

$$\left(18 - 15\frac{3}{8}\right)\text{ft} = \left(17\frac{8}{8} - 15\frac{3}{8}\right)\text{ft} = 2\frac{5}{8}\text{ft}$$

CHAPTER SIX
RATIONAL NUMBERS: DECIMALS

Section 6.1
Reading, Writing, Rounding, and Inequalities

1. 0.12 = twelve hundredths

3. 0.267 = two hundred sixty-seven thousandths

5. 6.0004 = six and four ten-thousandths

7. eleven hundredths = 0.11

9. one hundred eleven thousandths = 0.111

11. two and nineteen thousandths = 2.019

13. 0.504 = five hundred four thousandths

15. 50.04 = fifty and four hundredths

17. 18.0205 = eighteen and two hundred five ten-thousandths

19. twelve thousandths = 0.012

21. seven hundred and ninety-six thousandths = 700.096

23. five hundred five and five thousandths = 505.005

25. $0.33 = \dfrac{33}{100}$ 　　　　　　　27. $0.75 = \dfrac{75}{100} = \dfrac{3}{4}$

29. one hundred eleven thousandths = $\dfrac{111}{1000}$

31. $0.34 = \dfrac{34}{100} = \dfrac{17}{50}$ 　　　　33. $0.486 = \dfrac{486}{1000} = \dfrac{243}{500}$

35. two hundred thousandths = 0.200 = 0.2 = $\dfrac{2}{10} = \dfrac{1}{5}$

37. 0.1　　0.6　　0.7

39. 0.05　0.07　0.6

41. 4.159　　4.16　4.161

43. 0.072　　　　0.0729　　　0.073　　　0.073001　　0.073015

45. 0.88579　　　0.88799　　0.888　　　0.8881

47. 20.004　　　　20.039　　　20.04　　　20.093

49. The statement 3.1231 < 3.1213 is false

51. The statement 13.1204 < 13.2014 is true

53. Rounding Table:

		Unit	Tenth	Hundredth
53.	15.888	16	15.9	15.89
55.	477.774	478	477.8	477.77
57.	0.7392	1	0.7	0.74

61. $33.5374 rounded to the nearest cent is $33.54

65. $246.4936 rounded to the nearest cent is $246.49

67. Rounding Table:

	Number	Ten	Hundredth	Thousandths
63.	12.5532	10	12.55	12.553
65.	245.2454	250	245.25	245.245
67.	0.5536	0	0.55	0.554

69. $10.78 rounded to the nearest dollar is $11

71. $1129.38 rounded to the nearest dollar is $112

73. The number of inches of precipitation during the second week is one and three tenths inches

75. The number of inches of precipitation in the first week is $\dfrac{6}{100} = \dfrac{3}{50}$ inch.

77. The amount of precipitation in week four rounded to the nearest tenth of an inch is ____ inches.

79. Week one has the least precipitation.   Week two has the most precipitation.

81. \$64.79 = sixty-four and seventy-nine hundredths dollars

83. $0.125 = \dfrac{125}{1000} = \dfrac{1}{8}$

85. The position of the arrow to the nearest hundredth is 1.62.

87. The position of the arrow to the nearest tenth is 1.6.

89. The best bid is made by Circle K Meats for 98.35 cents per pound.

91. \$1617.37099921 rounded to the nearest cent is \$1617.37.

93. 127.659 rounded to the nearest tenth is 127.7 mi/hr.

95. Rounding 622.407 to 622.5 assumes that the digit to the right of the 4 is 5 or greater.   Correct roundings would be 622.41 or 622.4.

97. answers vary   99. answers vary

101. a. $7.44 < 7\dfrac{7}{18}$ is a false statement because $7.44 < 7.39$

   b. $8.6 > 8\dfrac{5}{9}$ is a true statement because $8.6 > 8.55$

   c. $3\dfrac{2}{7} < 3.285$ is a false statement because $3.2857... > 3.285$

   d. $9\dfrac{3}{11} > 9.271$ is a true statement because $9.2727 > 9.271$

103. answers vary   105. answers vary

107. $\dfrac{5}{6} + \left(-\dfrac{7}{8}\right) = \dfrac{5}{6} \cdot \dfrac{4}{4} + \left(-\dfrac{7}{8}\right) \cdot \dfrac{3}{3} = \dfrac{20}{24} + \left(-\dfrac{21}{24}\right) = -\dfrac{1}{24}$

109. The sum is:

$$\left(-\frac{1}{2}\right)+\left(-\frac{2}{3}\right)+\left(-\frac{3}{4}\right)=\left(-\frac{1}{2}\right)\left(\frac{6}{6}\right)+\left(-\frac{2}{3}\right)\left(\frac{4}{4}\right)+\left(-\frac{3}{4}\right)\left(\frac{3}{3}\right)=$$

$$\left(-\frac{6}{12}\right)+\left(-\frac{8}{12}\right)+\left(-\frac{9}{12}\right)=-\frac{23}{12}$$

Section 6.2
Adding and Subtracting Rational Numbers (Decimals)

1. $0.4 + 0.3 = 0.7$

3. $2.5 + 1.3 = 3.8$

5. $1.4 + 2.1 + 4.2 = 7.7$

7. $23.3 + 4.13 = 27.43$

9. To add 4.5, 6.78, 9.342 and 23, first rewrite each with 3 decimal places.

11. $8.3 + 5.541 = 13.841$

13. $8.28 + 0.28 + 12.3 + 2.54 = 23.40$

15. $0.438 + 0.834 + 1.483 = 2.755$

17. $0.0017 + 1.007 + 7 + 1.071 = 9.0797$

19. $37.008 + 38.007 + 3.87 + 3.708 = 82.593$

21. $7.5 + 14.378 + 33.6583 = 55.5363$

23. $43.524 + 12.8 + 774.943 = 831.267$

25. $9.76 + 9.6 + 0.581 + 7.04 = 26.981$

27. $0.9 - 0.6 = 0.3$

29. $5.7 - 2.3 = 3.4$

31. $8.31 - 3.21 = 5.10$

33. $19.05 - 12.64 = 6.41$

35. $16.20 - 8.11 = 8.09$

37. $0.612 - 0.155 = 0.457$

39. $2.712 - 1.148 = 1.564$

41. $5.678 - 3.069 = 2.609$

43. $134.98 - 67.936 = 67.044$

45. $12.1 - 9.34 = 2.76$

47. $8.642 - 8.573 = 0.069$

49. $0.09 + 0.04 + 0.04 = 0.17$

51. $0.7 - 0.4 = 0.3$

53. $4 + 2 + 0 + 0 = 6$

55. $0.08 - 0.00 = 0.08$

57. $0.5 + 0.1 + 0.3 + 0.7 = 1.6$

59. $0.0764 - 0.03621 = 0.04019$

61. $0.0342 + 0.00687 + 0.057294 + 0.00843 = 0.106794$

63. $11.9867 - 27.3974 = -15.3207$

65. $0.067 + (-0.456) + (-0.0964) + 0.5321 + (-0.112) = 0.5991 - 0.6644 = -0.0653$

67. $0.0456 - 0.7834 - 0.456 + 0.3097 - 0.5067 = 0.3553 - 1.7461 = -1.3908$

69. $135.904 - (-34.651) - 78.45 = 92.105$

71. $3.17t + 3t + 0.5t + 0.8t = 7.47t$

73. $1.21x - 9.34x = -8.13x$

75. $-23.4a - (-9.5a) = -13.9a$

77. $6.788x - (-3.408) - (-5.009x) - 8.1 = 11.797x - 4.692$

79. $-34.98b - (-10.05c) - (-8.04b) + 3.2c = -26.94b + 13.25c$

81. $(-34.98x - 10.5y) + (18.04x + 47y) = -16.94x + 36.5y$

83. $(-5a^2 - 3.46a) - (0.34a + 1.87a^2) =$
$(-5a^2 - 3.46a) + (-0.34a - 1.87a^2) = -6.87a^2 - 3.8a$

85. $9.2 \text{ gal} + 11.9 \text{ gal} + 15.4 \text{ gal} + 12.6 \text{ gal} = 49.1 \text{ gal}$

87. $45.984 + 134.;6 + 98.992 + 89.56 + 102.774 = 471.91$ rounded to the nearest tenth is 471.9.

89. $12.16 \text{ sec} - 11.382 \text{ sec} = 0.778 \text{ sec}$

91. $9.35 \text{ sec} + 9.91 \text{ sec} + 10.04 \text{ sec} + 9.65 \text{ sec} = 38.95 \text{ sec}$

93. $52.78 \text{ sec} - (12.83 \text{ sec} + 13.22 \text{ sec} + 13.56 \text{ sec}) = 52.78 \text{ sec} - 39.61 \text{ sec} = 13.17 \text{ sec}$

95. \$28.5 billion − \$9.8 billion = \$18.7 billion

97. $2796 - ($254.87 + $152.32 + $155.40 + $35.61 + $82.45 + $35 + $134.45) =$
$2796 - $850.10 = 1945.90

99. $8.23\% - 7.88\% = 0.35\%$

101. $47.7 \text{ ft} + 26.8 \text{ ft} = 74.5 \text{ ft} \Rightarrow 75 \text{ ft}$

103. $8.34375 \text{ in} - (1.4375 \text{ in} + 0.3125 \text{ in}) = 8.34375 \text{ in} - 1.75 \text{ in} = 6.59375 \text{ in}$

105. answers vary   107. answers vary

109. Twenty-three 8.75's must be added to have a sum that is greater than 200.
$23(8.75) = 201.25$

111. $0.2, 0.19, 0.188, \underline{0.1877}, 0.18766, 0.187655$

113. $6.5625 - 4.99 = 1.5725$

115. answers vary

117. $\left(\dfrac{7}{8}\right)\left(-\dfrac{2}{3}\right) = -\dfrac{7}{12}$

119. $(4x)(-13x)(x^2) = -52x^4$

Section 6.3
Multiplying Rational Numbers (Decimals)

1. $0.4 \times 8 = 3.2$ 3. $1.5 \times 6 = 9.0$

5. $3 \times 0.09 = 0.27$ 7. $0.5 \times 0.4 = 0.2$

9. $0.03 \times 0.5 = 0.015$ 11. $0.16 \times (-0.4) = -0.064$

13. The number of decimal places in the product of 9.456 and 4.23 is 5.

15. $7.45 \times 0.002 = 0.0149$

17. $1.45 \times 4.6 = 6.67$

19. $7.84 \times 0.53 = 4.1552$

21. $0.346 \times 7.8 = 2.6988$

23. $(8.52)(-3.54) = -30.1608$

25. $(6.5)(-0.6)(0.03) = -0.117$

27. $(0.06x)(1.2x) = 0.072x^2$

29. $(7.5t^2)(-0.3t^2) = -2.25t^4$

31. $30(0.8) = 24$

33. $0.05(0.07) = 0.0035$

35. $20(0.08) = 1.6$

37. $0.09(0.02) = 0.0018$

39. $23.5 \times 0.47 = 11.045$

41. $0.356 \times 0.067 = 0.023852$

43. $0.0975 \times 3.92 = 0.3822$

45. $0.825 \times 0.0054 = 0.004455$

47. 20.7 gal + 20.4 gal + 19.3 gal + 18.9 gal + 18.4 gal = 97.7 gal

49. 18.4 gal × \$1.523 = \$28.0232 rounded to the nearest cent is \$28.02

51. Grant paid the least for his fill-up at \$1.393/gal.

53. $(98.67)(3.52) = 347.3184$

55. $(12.6)(760.02) = 9576.252$

57. $(9.86)(146.3)(14.83) = 21{,}392.54194$ rounded to the nearest thousandth
     is 21,392.542

59. $(-5.7)(0.57)(-5.07)(50.7) = 835.152201$ rounded to the nearest hundredth
     is 835.15.

61. $5.3b(4b + 3.1) = 21.2b^2 + 16.43b$

63. $-4.5y(2.1y - 0.9) = -9.45y^2 + 4.05y$

65. $0.4b^2(3b - 0.4c) = 1.2b^2 - 0.16b^2 c$

67. \$37.83(14.7 yd) = \$556.101 rounded to the nearest cent is \$556.10

69. 3(\$39.72) + 312(\$0.32) = \$119.16 + \$99.84 = \$219.00

71. Rent of a midsize car: 3($39.72) + $0.32(345) = $119.16 + $110.40 = $229.56
Rent of a compact car: 5($25.95) + $0.25(345) = $129.75 + $86.25 = $216.00
It cost less to rent a compact car by: $229.56 – $216 = $13.56

73. Store 1: $75 + 18($85.95) = $75 + $1547.1 = $1622.10
Store 2: $125 + 24($63.25) = $125 + $1518 = $1643
Store 3: $300 + 12($109.55) = $300 + $1314.60 = $1614.60
Store 3 is selling the refrigerator for the least total cost.

75. 18.875 ft(3.8 lb/ft) = 71.725 lb

77. 38.5(0.6)(2.9) = 66.99 points

79. 1970: 4(132 lb) = 528 lb
1980: 4(126.4 lb) = 505.6 lb
1990: 4(112.3 lb) = 449.2 lb
The consumption of red meat has declined over the two decades between 1970 and
   One reason is that people are more aware that eating red meat can
   increase cholesterol and therefore, increase the risks of heart disease.

81. The tax amount owed is: $97,700$\left(\dfrac{\$0.199}{\$100}\right)$ = $194.42

83. answers vary

85. 84 is the smallest whole number that you can multiply 0.66 by to get a product that
   is greater than 55.

87. 1.8   0.36   0.108   0.0432   0.0216

89. answers vary

91. $67,000 \div 10^2 = 670$

93. $23 \times 10^7 = 230,000,000$

Section 6.4
Multiplying and Dividing by Powers of Ten and Scientific Notation

1. $4.25 \div 10 = 0.425$

3. $(3.67)(100) = 367$

5. $(0.62833)(1000) = 628.33$

7. $\dfrac{569.2}{1000} = 0.5692$

9. $5645 \div 100 = 56.45$

11. $0.87 \times 10^4 = 8700$

13. To multiply 4.56 by 10^5 move the decimal point five places to the <u>right.</u>

15. $(6.274)(1000) = 6274$

17. $1.85 \div 10 = 0.185$

19. $36.9(1000) = 36,900$

21. $\dfrac{6895.3}{10,000} = 0.68953$

23. $14.78(100,000) = 1,478,000$

25. $1367.94 \div 100 = 13.6794$

27. $45.8 \div 100,000 = 0.000458$

29. $230,000 = 2.3 \times 10^5$

31. $0.00035 = 3.5 \times 10^{-4}$

33. $467.95 = 4.6795 \times 10^2$

35. $6 \times 10^4 = 60,000$

37. $8 \times 10^{-3} = 0.008$

39. $4.78 \times 10^3 = 4,780$

41. $780,000 = 7.8 \times 10^5$

43. $0.0000345 = 3.45 \times 10^{-5}$

45. $0.0000000000821 = 8.21 \times 10^{-11}$

47. $3567.003 = 3.567003 \times 10^3$

49. $1.345 \times 10^{-6} = 0.000001345$

51. $7.11 \times 10^9 = 7,110,000,000$

53. $4.44 \times 10^{-7} = 0.000000444$

55. $5.6723 \times 10^2 = 567.23$

57. $100(\$22.29) = \$2229.00$

59. $100(\$985) = \$98,500$

61. $52,000,000 = 5.2 \times 10^7$ sq. mi.

63. 0.366 is being multiplied by 1000

65. $424 \div 1000 = 0.424$

67. $0.0000033 = 3.3 \times 10^{-6}$

69. Mean Distance in Miles Table:

Planet	Mean Distance in Miles	Mean Distance in Miles
Mercury	3.6×10^7	36,000,000
Venus	6.724×10^7	67,240,000
Earth	9.296×10^7	92,960,000
Mars	1.4164×10^8	141,640,000
Jupiter	4.8364×10^8	483,640,000
Saturn	8.87×10^8	887,000,000
Uranus	1.783×10^9	1,783,000,000
Neptune	2.795×10^9	2,795,000,000
Pluto	3.666×10^9	3,666,000,000

71. 1.3×10^{-3} in $= 0.0013$ in

73. $100,000(15.5 \text{ lb}) = 1,550,000 \text{ lb} = 1.55 \times 10^6 \text{ lb of fish}$
    $100,000(63.6 \text{ lb}) = 6,360,000 \text{ lb} = 6.36 \times 10^6 \text{ lb of poultry}$
    $100,000(112.3 \text{ lb}) = 11,230,000 \text{ lb} = 1.123 \times 10^7 \text{ lb of red meat}$

75. $\$26,326,000,000 - \$24,744,000,000 = \$1,582,000,000 = \$1.582 \times 10^9$
    The average amount of increase per year $= \$1.582 \times 10^9 \div 10 = \$1.582 \times 10^8$

77. $206,265(93,000,000 \text{ mi.}) = 19,182,645,000,000 \text{ mi} = 1.92 \times 10^{13} \text{ mi}$

79. $\dfrac{\left(3.25 \times 10^{-3}\right)\left(2.4 \times 10^3\right)}{\left(4.8 \times 10^{-4}\right)\left(2.5 \times 10^{-3}\right)} = \dfrac{7.8 \times 10^0}{12 \times 10^{-7}} = 0.65 \times 10^7 = 6.5 \times 10^6$

81. answers vary

83. $8208 \div 76 = 108$

85. $15,023 \div 25 = 600$ r 23

Section 6.5
Dividing Rational Numbers (Decimals)

1. $3.5 \div 7 = 0.5$

3. $19.6 \div 2 = 9.8$

5. $18.31 \div 0.1 = 183.1$

7. $242.4 \div 0.12 = 2020$

9. To divide 2.65 by 0.05 we first multiply both the dividend and the divisor by 100 so we are dividing by a <u>whole number.</u>

11. $1008 \div 80 = 12.6$

13. $-16.64 \div 32 = -0.52$

15. $8.96 \div 7 = 1.28 \Rightarrow 1.3$

17. $11.778 \div 1.3 = 9.06 \Rightarrow 9.1$

19. $34.22 \div 2.2 = 15.5545 \Rightarrow 15.555$

21. $\dfrac{3}{4} = 0.75$

23. $\dfrac{3}{8} = 0.375$

25. $\dfrac{11}{16} = 0.6875$

27. $3\dfrac{11}{20} = 3.55$

29. $11\dfrac{3}{125} = 11.024$

31. Convert fractions to decimals:

		Tenth	Hundredth
31.	$\dfrac{3}{7}$	0.4	0.43
33.	$\dfrac{5}{11}$	0.5	0.45
35.	$\dfrac{2}{13}$	0.2	0.15
37.	$\dfrac{8}{15}$	0.5	0.53
39.	$5\dfrac{17}{18}$	5.9	5.94

41. $0.08 \div 4 = 0.02$

43. $2.6 \div 0.2 = 13$

45. $0.5 \div 0.5 = 1.0$

47. $32 \div 0.08 = 400$

49. $6211.84 \div 64 = 97.06$

51. $0.2267 \div 2.97 = 0.076 \Rightarrow 0.08$

53. $0.12 \div 0.007 = 17.142 \Rightarrow 17.14$

55. $\dfrac{8.4x}{-0.4} = -21x$

57. $\dfrac{-0.45m}{1.2} = -0.375m$

59. $\dfrac{-9.19}{0.11} = -83.545 \Rightarrow -83.55$

61. $\$4.42 \div 3 = \$1.4733/\text{lb} \Rightarrow \$1.473/\text{lb}$

63. $\$9.62 \div 3.24 = \$2.9691/\text{lb} \Rightarrow \$2.969/\text{lb}$

65. $7(\$16.05 \div 4.6) = 7(\$3.489) = \$24.423 \Rightarrow \$24.42$

67. $6(\$2.969) + 2(\$0.795) + 4(\$0.077) + 5(\$1.473) =$
    $\$17.841 + \$1.59 + \$0.308 + \$7.365 = \$27.077 \Rightarrow \$27.08$

69. $\dfrac{3.8a - 5.7}{1.9} = 2a - 3$

71. $\dfrac{-2.16a^2 - 0.0096a}{0.24a} = -9a - 0.04$

73. The rounding table is:

		Hundredth	Thousandth
73.	$\dfrac{12}{31}$	0.39	0.387
75.	$\dfrac{85}{91}$	0.93	0.934

77. $\dfrac{11}{12} = 0.91\overline{6}$

79. $\dfrac{23}{26} = 0.8\overline{846153}$

81. $\$12.45 \div 3 = \$4.15$

83. $65 \text{ gal} \div 1.94 \text{ gal/day} = 33.50 \text{ days} \Rightarrow 34 \text{ days}$

85. $\$6355.48 \div 4 = \$1588.87$

87. $630.6 \text{ lb} \div 26.16 \text{ lb/ft} \approx 24.1 \text{ ft}$

89. $2(\$312.6 \div 56) = 2(\$5.582) = \$11.16$

91. $\text{ERA} = \dfrac{26}{80 \div 9} = \dfrac{26}{8.888} = 2.925 \Rightarrow 2.93$

93. $\dfrac{18}{29} = 0.6206 \Rightarrow 0.621$

95. answers vary

97. answers vary

99. $2.5 \times 10^{-4} < \dfrac{3}{2000}$ because $0.00025 < 0.0$

101. answers vary

103. $2\dfrac{1}{4} - \dfrac{3}{4} + 1\dfrac{5}{8} = 2\dfrac{2}{8} - \dfrac{6}{8} + 1\dfrac{5}{8} = 3\dfrac{7}{8} - \dfrac{6}{8} = 3\dfrac{1}{8}$

105. $17 - 5\dfrac{7}{9} = 16 + \dfrac{9}{9} - 5\dfrac{7}{9} = 11\dfrac{2}{9}$

107. Mr. Lewis buys 350 books for $60.

He sells $\dfrac{2}{5}(350) = 140$ books for $25.   There are 210 books left.

He sells 26 books at $1.50(26) = $39.   There are 184 books left.
He sells 45 books at $1.00(45) = $45.   There are 139 books left to give away.
He spent $60 + $15 = $75.   He took in $25 + $39 + $45 = $109.
His profit is $109 − $75 = $34

Section 6.6
Another Look at Conversion of Units

1. $8\,\text{oz} \cdot \dfrac{\text{lb}}{16\,\text{oz}} = 0.5\,\text{lb}$

3. $15\,\text{mm} \cdot \dfrac{\text{cm}}{10\,\text{mm}} = 1.5\,\text{cm}$

5. $550\,\text{m}l \cdot \dfrac{l}{1000\,\text{m}l} = 0.55\,l$

7. $6.9\,\text{ft} \cdot \dfrac{\text{yd}}{3\,\text{ft}} = 2.3\,\text{yd}$

9. $3.02\,\text{ft} \cdot \dfrac{12\,\text{in}}{\text{ft}} - 36.24\,\text{in}$

11. $1.83\,\text{m} \cdot \dfrac{1000\,\text{mm}}{\text{m}} = 1830\,\text{mm}$

13. $4.56\,\text{mi} \cdot \dfrac{5280\,\text{ft}}{\text{mi}} = 24{,}076.8\,\text{ft}$

15. $16\,\text{in} \cdot \dfrac{2.54\,\text{cm}}{\text{in}} = 40.64\,\text{cm}$

17. $4.5\,l \cdot \dfrac{1.057\,\text{qt}}{l} = 4.76\,\text{qt}$

19. $37.9\,\text{cm} \cdot \dfrac{\text{in}}{2.54\,\text{cm}} = 14.92\,\text{in}$

21. $9.5 \text{ sq in} \cdot \dfrac{\text{sq.ft.}}{144 \text{ sq.in}} = 0.07 \text{ sq.ft.}$

23. $53,400 \text{ cm}^2 \cdot \dfrac{\text{m}^2}{100 \text{ cm}^2} = 5.34 \text{ m}^2$

25. $14.8 \text{ ft} \cdot \dfrac{0.3048 \text{ m}}{\text{ft}} = 4.5 \text{ m}$

27. $3.2 \text{ kg} \cdot \dfrac{\text{lb}}{0.454 \text{ kg}} = 7.0 \text{ lb}$

29. $8235 \text{ m} \cdot \dfrac{\text{km}}{1000 \text{ m}} \dfrac{0.6214 \text{ mi}}{\text{km}} = 5.1 \text{ mi}$

31. $7.5 \text{ c} \cdot \dfrac{\text{qt}}{4 \text{ c}} \dfrac{0.946 l}{\text{qt}} \cdot \dfrac{100 cl}{l} = 177.4 \, cl$

33. $14.5 \text{ ft}^2 \cdot \dfrac{\text{m}^2}{(3.281)^2 \text{ ft}^2} = 14.5 \text{ ft}^2 \cdot \dfrac{\text{m}^2}{10.76 \text{ ft}^2} = 1.3 \text{ m}^2$

35. $\dfrac{1.5 \text{ lb}}{\text{ft}} \cdot \dfrac{3.281 \text{ ft}}{\text{m}} \cdot \dfrac{\text{g}}{0.0022 \text{ lb}} = 2237.05 \text{ g/m}$

37. $\dfrac{5.2 \text{ lb}}{\text{ft}^2} \cdot \dfrac{\text{ft}^2}{(0.3048)^2 \text{ m}^2} \cdot \dfrac{\text{m}^2}{100^2 \text{ cm}^2} \cdot \dfrac{453.59 \text{ g}}{\text{lb}} = 2.54 \text{ g/cm}^2$

39. $\dfrac{525 \text{ g}}{l} \cdot \dfrac{\text{lb}}{453.59 \text{ g}} \cdot \dfrac{l}{1.057 \text{ qt}} = 1.10 \text{ lb/qt}$

41. $18.9 \text{ oz} \cdot \dfrac{\text{lb}}{16 \text{ oz}} \cdot \dfrac{453.59 \text{ g}}{\text{lb}} = 535.8 \text{ g}$

43. $\dfrac{30 \text{ cents}}{\text{mi}} \cdot \dfrac{\$}{100 \text{ cents}} \cdot \dfrac{\text{mi}}{1.609 \text{ km}} = \$0.19 \text{ km}$

45. $\dfrac{295\,\text{lb}}{1} \cdot \dfrac{0.454\,\text{kg}}{\text{lb}} = 133.9\,\text{kg}$

47. $\dfrac{95\,\text{ft}}{1} \cdot \dfrac{0.3048\,\text{m}}{\text{ft}} = 29\,\text{m}$

49. United States: $\dfrac{6\,\text{lb}}{1} \cdot \dfrac{453.59\,\text{g}}{\text{lb}} \cdot \dfrac{\text{kg}}{1000\,\text{g}} = 2.7\,\text{kg} \quad \Rightarrow \quad \dfrac{\$17.95}{2.7\,\text{kg}} = \$6.65/\text{kg}$

   Canada: $\dfrac{\$18.95}{3\,\text{kg}} = \$6.31/\text{kg}$

   Therefore, the coffee sold in Canada is the better buy.

51. $\dfrac{26.83\,\text{in}}{1} \cdot \dfrac{2.54\,\text{cm}}{\text{in}} = 68.1\,\text{cm}$

53. $\dfrac{1\,\text{min }10\,\text{sec}}{100\,\text{yd}} = \dfrac{70\,\text{sec}}{100\,\text{yd}} = 0.7\,\text{sec/yd}$ $\dfrac{0.7\,\text{sec}}{\text{yd}} \cdot \dfrac{1.094\,\text{yd}}{\text{m}} = 0.7658\,\text{yd/m}$

55. Doug's expected time for the 100-m breaststroke is 1 min 18.7 sec

57. $\dfrac{1\,\text{nautical mile}}{1} \cdot \dfrac{6076\,\text{ft}}{\text{nautical mile}} \cdot \dfrac{\text{m}}{3.281\,\text{ft}} \cdot \dfrac{\text{km}}{1000\,\text{m}} = 1.852\,\text{km}$

59. $\dfrac{1\,\text{mi}}{1} \cdot \dfrac{5280\,\text{ft}}{\text{mi}} \cdot \dfrac{\text{nautical mile}}{6076\,\text{ft}} = 0.869\,\text{nautical miles}$

61. answers vary

63. answers vary

65. $5\,\text{gal} \cdot \dfrac{4\,\text{qt}}{\text{gal}} + 3\,\text{qt} + 1\,\text{pt} \cdot \dfrac{\text{qt}}{2\,\text{pt}} = 20\,\text{qt} + 3\,\text{qt} + \dfrac{1}{2}\,\text{qt} =$

   $23\dfrac{1}{2}\,\text{qt} \cdot \dfrac{l}{1.057\,\text{qt}} = \dfrac{47}{2}\,\text{qt} \cdot \dfrac{l}{1.057\,\text{qt}} = 22.2\,l$

67. group activity

69. Evaluate:

$$2xyz - 3xz$$

$$2\left(\frac{7}{12}\right)\left(\frac{5}{8}\right)\left(\frac{3}{5}\right) - 3\left(\frac{7}{12}\right)\left(\frac{3}{5}\right)$$

$$\frac{7}{16} - \frac{21}{20}$$

$$\frac{7}{16} \cdot \frac{5}{5} - \frac{21}{20} \cdot \frac{4}{4}$$

$$\frac{35}{80} - \frac{84}{80}$$

$$-\frac{49}{80}$$

71. $x = -\frac{2}{3}$ is not a solution

$$2x^2 + 17x + 8 = 0$$

$$2\left(-\frac{2}{3}\right)^2 + 17\left(-\frac{2}{3}\right) + 8 = 0$$

$$2\left(\frac{4}{9}\right) + 17\left(-\frac{2}{3}\right) + 8 = 0$$

$$\frac{8}{9} - \frac{34}{3} + 8 = 0$$

$$\frac{8}{9} - \frac{34}{3} \cdot \frac{3}{3} + \frac{8}{1} \cdot \frac{9}{9} = 0$$

$$\frac{8}{9} - \frac{102}{9} + \frac{72}{9} = 0$$

$$-\frac{22}{9} \neq 0$$

Section 6.7
Evaluating Algebraic Expressions and Formulas
With Rational Numbers (Decimals)

1. $ab = (0.5)(-3) = -1.5$

3. $abx = (0.6)(3)(0.01) = 0.018$

5. $b - y = -3.56 - 7.11 = -10.67$

7. $x^3 = (0.2)^3 = 0.008$

9. $\dfrac{a}{b} = \dfrac{-0.44}{-1.1} = 0.4$

11. 40.4 is not a solution
$$x + 3.3 \neq 43.73$$
$$40.4 + 3.3 \neq 43.73$$
$$43.7 \neq 43.73$$

13. Evaluate:
$$a^2 x^2$$
$$(-0.3)^2 (2.02)^2$$
$$(0.09)(4.0804)$$
$$0.367236$$

15. Evaluate:
$$ab + yz$$
$$(-0.3)(100) + (0.4)(-0.45)$$
$$-30 + (-0.18)$$
$$-30.18$$

17. Evaluate:
$$bc - by$$
$$100(-0.26) - 100(0.4)$$
$$-26 - 40$$
$$-66$$

19. Evaluate:
$$cbx - ay$$
$$(-0.26)(100)(2.02) - (-0.3)(0.4)$$
$$-52.52 + 0.12$$
$$-52.4$$

21. Evaluate:
$$b^2(a + y)$$
$$(100)^2(-0.3 + 0.4)$$
$$10,000(0.1)$$
$$1000$$

23. Evaluate:
$$(a + x)(b + y)$$
$$(-0.3 + 2.02)(100 + 0.4)$$
$$1.72(100.04)$$
$$172.688$$

25. 0.283 is a solution
$$4x + 3 = 4.132$$
$$4(0.283) + 3 = 4.132$$
$$1.132 + 3 = 4.132$$
$$4.132 = 4.132$$

27. $a = \dfrac{bh}{2} = \dfrac{6.5(3.9)}{2} = \dfrac{25.35}{2} = 12.675$

29. $V = s^3 = (3.56)^3 = 45.11 \Rightarrow 45.1$

31. $\dfrac{a+c}{b} = \dfrac{0.15 + (-0.342)}{55} = \dfrac{-0.192}{55} = -0.003$

33. $\dfrac{(a+b)^2}{b} = \dfrac{(0.15 + 55)^2}{55} = \dfrac{(55.15)^2}{55} = \dfrac{3041.5225}{55} = 55.300$

35. Evaluate:

$$5x^2 - ac$$

$$5(-4.53)^2 - (0.15)(-0.342)$$

$$5(20.5209) + 0.0513$$

$$102.6045 + 0.0513$$

$$102.6558 \Rightarrow 102.656$$

37. 2.3 is a solution

$$3w - 7 = -0.1$$

$$3(2.3) - 7 = -0.1$$

$$6.9 - 7 = -0.1$$

$$-0.1 = -0.1$$

39. 0.34 is not a solution

$$7.64 - 2m \neq 0.68$$

$$7.64 - 2(0.34) \neq 0.68$$

$$7.64 - 0.68 \neq 0.68$$

$$6.96 \neq 0.68$$

41. $d = rt = (63.45 \text{ mi/hr})\ 4.5 \text{ hr} = 285.525 \text{ mi}$

43. The area is:

$$A = \frac{\pi r^2}{2} = \frac{3.14(125 \text{ ft})^2}{2} = \frac{3.14(15,625 \text{ ft}^2)}{2} = \frac{49062.5 \text{ ft}^2}{2}$$

$$= 24,531.25 \text{ ft}^2 = 24,531 \text{ ft}^2$$

45. The commission is:

$$C = \$2000 + 0.03(h - \$40,000)$$

$$C = \$2000 + 0.03(\$135,850 - \$40,000)$$

$$C = \$2000 + 0.03(\$95,850)$$

$$C = \$2000 + \$2875.5$$

$$C = \$4875.50$$

47. The surface area of the cylindrical drum is:

$$A = 2\pi r h + 2\pi r^2$$

$$A = 2(3.14)(7.85 \text{ ft})(10.5 \text{ ft}) + 2(3.14)(7.85 \text{ ft})^2$$

$$A = 517.6 \text{ ft}^2 + 2(3.14)(61.62 \text{ ft}^2)$$

$$A = 517.6 \text{ ft}^2 + 386.97 \text{ ft}^2$$

$$A = 904.6 \text{ ft}^2$$

The number of gallons of paint that must be purchased is:

$$904.6 \text{ ft}^2 \div 125 \text{ ft}^2 / \text{gal} = 7.237 \text{ gal} \Rightarrow 8 \text{ gal}$$

49. $T = 3(34) + 33 = 102 + 33 = 135 \text{ pt}$

51. The average number of farms lost between 1920 and 1950 was:
   6.0 million − 5.5 million = 0.5 million = 500,000 farms
   The average number of farms lost per year between 1920 and 1950 was:
   500,000 farms ÷ 30 years = 16,667 farms ⇒ 17,000 farms/year

53. Cyclist Distance Table:

Time (hours)	1	2	3	4	5
Total Distance of Fastest Cyclist	19	38	57	76	95
Total Distance of Slowest Cyclist	16	32	48	64	80

55. Fastest cyclist: 150 mi ÷ 19 mi/hr = 7.89 hr
   6:30 AM + 7.89 hr = 6:30 AM + 7 hr + 60 min(0.89) =
   6:30 AM + 7 hr 54 min = 2:24 PM
   Slowest cyclist: 150 mi ÷ 16 mi/hr = 9.38 hr
   6:30 AM + 9.38 hr = 6:30 AM + 9 hr + 60 min(0.38) =
   6:30 AM + 9 hr 23 min = 3:53 PM

57. horsepower $= \dfrac{\text{PLAN}}{33{,}000} = \dfrac{(75)(2.5)(30.05)(75)}{33{,}000} = \dfrac{422{,}578.125}{33{,}000} = 12.8$

59. The area of the shaded ring is:
   $$A = \pi R^2 - \pi r^2$$
   $$A = 3.14(4.5\,\text{in})^2 - 3.14(2.5\,\text{in})^2$$
   $$A = 3.14(20.25\,\text{in}^2) - 3.14(6.25\,\text{in}^2)$$
   $$A = 63.59\,\text{in}^2 - 19.63\,\text{in}^2$$
   $$A = 43.96\,\text{in}^2$$

61. Solve:
   $$5x + 33 = 13$$
   $$5x = -20$$
   $$x = -4$$

63. Solve:
   $$3 - 5y = 63$$
   $$-5y = 60$$
   $$y = -12$$

65. Solve:
   $$\frac{3}{8}b + \frac{6}{5} = \frac{3}{2}$$
   $$40 \cdot \frac{3}{8}b + 40 \cdot \frac{6}{5} = 40 \cdot \frac{3}{2}$$
   $$15b + 48 = 60$$
   $$15b = 12$$
   $$b = \frac{12}{15} = \frac{4}{5}$$

Section 6.8
Solving Equations Involving Rational Numbers (Decimals)

1. Solve:
$$x + 4.52 = 8.93$$
$$x + 4.52 - 4.52 = 8.93 - 4.52$$
$$x = 4.41$$

3. Solve:
$$a - 0.005 = 5.342$$
$$a - 0.005 + 0.005 = 5.342 + 0.005$$
$$a = 5.347$$

5. Solve:
$$3x = -9.036$$
$$\frac{3x}{3} = \frac{-9.036}{3}$$
$$x = -3.012$$

7. Solve:
$$c - (-4.75) = -5.43$$
$$c + 4.75 = -5.43$$
$$c + 4.75 - 4.75 = -5.43 - 4.75$$
$$c = -10.18$$

9. Solve:
$$-6b = -36.066$$
$$\frac{-6b}{-6} = \frac{-36.066}{-6}$$
$$b = 6.011$$

11. Solve:
$$0.02c = 3.4$$
$$\frac{0.02c}{0.02} = \frac{3.4}{0.02}$$
$$c = 170$$

13. Solve:
$$4b - 0.66 = 32.34$$
$$4b - 0.66 + 0.66 = 32.34 + 0.66$$
$$4b = 33$$
$$\frac{4b}{4} = \frac{33}{4}$$
$$b = 8.25$$

15. Solve:
$$0.7c - 22.6 = 16.6$$
$$0.7c - 22.6 + 22.6 = 16.6 + 22.6$$
$$0.7c = 39.2$$
$$\frac{0.7c}{0.7} = \frac{39.2}{0.7}$$
$$c = 56$$

17. Solve:
$$-0.11x = 0.3872$$
$$\frac{-0.11x}{-0.11} = \frac{0.3872}{-0.11}$$
$$x = -3.52$$

19. Solve:
$$y - (-24) = 15.67$$
$$y + 24 = 15.67$$
$$y + 24 - 24 = 15.67 - 24$$
$$y = -8.33$$

21. Solve:
$$w + 5.67 = -1.567$$
$$w + 5.67 - 5.67 = -1.567 - 5.67$$
$$w = -7.237$$

23. Solve:
$$0.2y - 34.78 = 15.43$$
$$0.2y - 34.78 + 34.78 = 15.43 + 34.78$$
$$0.2y = 50.21$$
$$\frac{0.2y}{0.2} = \frac{50.21}{0.2}$$
$$y = 251.05$$

25. Solve:
$$4.8b - 0.83 = -9.95$$
$$4.8b - 0.83 + 0.83 = -9.95 + 0.83$$
$$4.8b = -9.12$$
$$\frac{4.8b}{4.8} = \frac{-9.12}{4.8}$$
$$b = -1.9$$

27. Solve:
$$8.7 = 7.8c + 13.77$$
$$8.7 - 13.77 = 7.8c + 13.77 - 13.77$$
$$-5.07 = 7.8c$$
$$\frac{-5.07}{7.8} = \frac{7.8c}{7.8}$$
$$-0.65 = c$$

29. The balance is:
$$D = B - nP$$
$$D = \$785 - 7(\$45.95)$$
$$D = \$785 - \$321.65$$
$$D = \$463.35$$

31. The number of payments needed is:
$$D = B - nP$$
$$\$0 = \$7100 - n(\$258.67)$$
$$-\$7100 = -n(\$258.67)$$
$$\frac{-\$7100}{-\$258.67} = \frac{-n(\$258.67)}{-\$258.67}$$
$$27 = n$$
plus the final payment = 28 payments

The amount of the final payment is:

$$F = \$7100 - 27(\$258.67)$$
$$F = \$7100 - \$6984.09$$
$$F = \$115.91$$

33. The amount of interest paid is:
$$D = B - nP$$
$$\$0 = \$125,000 + i - 30(12)(\$931.68)$$
$$\$0 = \$125,000 + i - \$335,404.80$$
$$\$0 = i - \$210,404.80$$
$$i = \$210,404.80$$

35. Solve:
$$78.43 + 6.81t = -4.56$$
$$6.81t = -82.99$$
$$t = -12.18$$
$$t \approx -12.2$$

37. Solve:
$$5.7a - 9.54 - 3.23a = 4.872$$
$$2.47a = 14.412$$
$$a = 5.83$$
$$a \approx 5.8$$

39. Solve:
$$55.78 = 9.83t - 34.78 - 13.59t$$
$$90.56 = -3.76t$$
$$-24.08 = t$$
$$-24.1 \approx t$$

41. Solve:
$$0.0356 - 0.0236z = 0.0976z + 0.00421$$
$$0.03139 = 0.1212z$$
$$0.25 = z$$
$$0.3 \approx z$$

43. The length is:
$$P = 2l + 2w$$
$$48.96 \text{ m} = 2l + 2(5.62 \text{ m})$$
$$48.96 \text{ m} = 2l + 11.24 \text{ m}$$
$$37.72 \text{ m} = 2w$$
$$18.86 \text{ m} = w$$

45. The initial velocity is:
$$v_1 = v_0 + 32.2t$$
$$305.5 = v_0 + 32.2(8.2)$$
$$305.5 = v_0 + 264.04$$
$$41.46 \text{ ft/sec} = v_0$$

47. The base is:
$$A = 0.5bh$$
$$243.6 \text{ cm}^2 = 0.5b(12.6 \text{ cm})$$
$$243.6 \text{ cm}^2 = 6.3 \text{ cm} \cdot b$$
$$38.666 \text{cm} = b$$
$$38.67 \text{ cm} \approx b$$

49. The player would need 185 bases.
$$651 = \frac{1000B}{285}$$
$$0.651 = \frac{B}{285}$$
$$185.5 \approx B$$

51. $21{,}998 \div 32.2 = 683.17 \approx 684 \text{ games}$

53. The Celsius temperature is:
$$5(32) = 9C + 160$$
$$160 = 9C + 160$$
$$0 = 9C$$
$$0° = C$$

55. Solve:
$$0.002x - 0.73 + 1.008x = 4.825$$
$$1.01x - 0.73 = 4.825$$
$$1.01x = 5.555$$
$$x = 5.5$$

57. Solve by eliminating the decimal fractions from the equation:

$$1.2a - 0.08 = 0.624a + 0.424$$

$$1000(1.2a - 0.08) = 1000(0.624a + 0.424)$$

$$1200a - 80 = 624a + 424$$

$$576a = 504$$

$$a = 0.875$$

59. Solve:

$$3.2(8.1 - 7.55) - 8.2x + 3(0.125 + 0.225) = 32.33$$

$$3.2(0.55) - 8.2x + 3(0.35) = 32.33$$

$$1.76 - 8.2x + 1.05 = 32.33$$

$$2.81 - 8.2x = 32.33$$

$$-8.2x = 29.52$$

$$x = -3.6$$

61. Solve:

$$0.43(1.02y - 6.7) - 0.56(0.12y - 3.45) = 9.843$$

$$0.4386y - 2.881 - 0.0672y + 1.932 = 9.843$$

$$0.3714y - 0.949 = 9.843$$

$$0.3714y = 10.792$$

$$y = 29.05$$

$$y \approx 29.1$$

63. answers vary

65. Evaluate:

$$a^2 + d^2$$

$$(-5)^2 + (-6)^2$$

$$25 + 36$$

$$61$$

67. Evaluate:

$$a^2 - d^2$$

$$\left(\frac{1}{2}\right)^2 - \left(-\frac{5}{6}\right)^2$$

$$\frac{1}{4} - \frac{25}{36}$$

$$\frac{1}{4} \cdot \frac{9}{9} - \frac{25}{36} = \frac{9}{36} - \frac{25}{36}$$

$$-\frac{16}{36} = -\frac{4}{9}$$

Section 6.9
Square Roots and the Pythagorean Formula

1. $\sqrt{81} = 9$ 3. $\sqrt{121} = 11$ 5. $\sqrt{\dfrac{4}{25}} = \dfrac{2}{5}$ 7. $\sqrt{\dfrac{121}{144}} = \dfrac{11}{12}$

9. $\sqrt{484} = 22$ 11. $\sqrt{116} \approx 10.77 \approx 10.8$

13. $-\sqrt{78} \approx -8.831 \approx -8.83$ 15. $\sqrt{21} \approx 4.58 \approx 4.6$

17. $\sqrt{0.675} \approx 0.821 \approx 0.82$

19. Solve:
$$a^2 + b^2 = c^2$$
$$a^2 + 8^2 = 17^2$$
$$a^2 + 64 = 289$$
$$a^2 = 225$$
$$a = 15$$

21. Solve:
$$a^2 + b^2 = c^2$$
$$12^2 + 5^2 = c^2$$
$$144 + 25 = c^2$$
$$169 = c^2$$
$$13 = c$$

23. Solve:
$$a^2 + b^2 = c^2$$
$$a^2 + 16^2 = 20^2$$
$$a^2 + 256 = 400$$
$$a^2 = 144$$
$$a = 12$$

25. Solve:
$$a^2 + b^2 = c^2$$
$$40^2 + 30^2 = c^2$$
$$1600 + 900 = c^2$$
$$2500 = c^2$$
$$50 = c$$

27. Solve:
$$a^2 + b^2 = c^2$$
$$2^2 + 3^2 = c^2$$
$$4 + 9 = c^2$$
$$13 = c^2$$
$$3.61 \approx c$$

29. Solve:
$$a^2 + b^2 = c^2$$
$$5^2 + b^2 = 7.7^2$$
$$25 + b^2 = 59.29$$
$$b^2 = 34.29$$
$$b \approx 5.86$$

31. Solve:
$$a^2 + b^2 = c^2$$
$$110^2 + b^2 = 175^2$$
$$12{,}100 + b^2 = 30{,}625$$
$$b^2 = 18{,}525$$
$$b \approx 136.11$$

33. Solve:
$$a^2 + b^2 = c^2$$
$$5.7^2 + 13.2^2 = c^2$$
$$32.49 + 174.24 = c^2$$
$$206.73 = c^2$$
$$14.38 \approx c$$

35. Solve:
$$a^2 + b^2 = c^2$$
$$37^2 + 55^2 = c^2$$
$$1369 + 3025 = c^2$$
$$4394 = c^2$$
$$66.29 \approx c$$

37. This is a right triangle.
$$a^2 + b^2 = c^2$$
$$16^2 + 30^2 = 34^2$$
$$256 + 900 = 1156$$
$$1156 = 1156$$

39. This is not a right triangle.
$$a^2 + b^2 = c^2$$
$$9^2 + 12^2 = 16^2$$
$$81 + 144 = 256$$
$$225 \neq 256$$

41. This is a right triangle.
$$a^2 + b^2 = c^2$$
$$3.2^2 + 2.4^2 = 4^2$$
$$10.24 + 5.76 = 16$$
$$16 = 16$$

43. This is not a right triangle
$$a^2 + b^2 = c^2$$
$$3.5^2 + 8.4^2 = 9.2^2$$
$$12.25 + 70.56 = 84.64$$
$$82.81 \neq 84.64$$

45. $t = \sqrt{\dfrac{d}{16}} = \sqrt{\dfrac{50}{16}} = \dfrac{\sqrt{50}}{4} \approx \dfrac{7.07}{4} \approx 1.76 \approx 1.8 \text{ sec}$

47. The length of the third side is:
$$a^2 + b^2 = c^2$$
$$25^2 + 39^2 = c^2$$
$$625 + 1521 = c^2$$
$$2146 = c^2$$
$$46.3 \text{ ft} \approx c$$

49. The completed chart is:

		Nearest Tenth	Nearest Hundredth	Nearest Thousandth	Nearest Ten-thousandth
49.	$\sqrt{365.96}$	19.1	19.13	19.130	19.1301
51.	$\sqrt{20.037}$	4.5	4.48	4.476	4.4763

53. $t = 2\pi\sqrt{\dfrac{l}{g}} = 2 \cdot \dfrac{22}{7} \cdot \sqrt{\dfrac{49}{36}} = \dfrac{44}{7} \cdot \dfrac{7}{6} = \dfrac{22}{3} = 7\dfrac{1}{3}$

55. $r \approx 0.564\sqrt{A} = 0.564\sqrt{706} = 0.564(26.57) = 14.98 \approx 15.0 \, \text{in}$

57. The length of the rafter is:
$$a^2 + b^2 = c^2$$
$$6^2 + 12^2 = c^2$$
$$36 + 144 = c^2$$
$$180 = c^2$$
$$13.42 \, \text{ft} \approx c$$

59. The length of the cable is:
$$a^2 + b^2 = c^2$$
$$35^2 + 50^2 = c^2$$
$$1225 + 2500 = c^2$$
$$3725 = c^2$$
$$61.0 \, \text{ft} \approx c$$

61. The plane flies:
$$a^2 + b^2 = c^2$$
$$200^2 + 50^2 = c^2$$
$$40,000 + 2500 = c^2$$
$$42,500 = c^2$$
$$206.2 \, \text{mi} \approx c$$

63. The distance is:
$$a^2 + b^2 = c^2$$
$$90^2 + 90^2 = c^2$$
$$8100 + 8100 = c^2$$
$$16,200 = c^2$$
$$127.3 \, \text{ft} \approx c$$

65. The shortest land distance is:
$$a^2 + b^2 = c^2$$
$$8.3^2 + 5.9^2 = c^2$$
$$68.89 + 34.81 = c^2$$
$$103.7 = c^2$$
$$10.2 \, \text{mi} \approx c$$

67. Answers vary

69. Simplify:

$$\sqrt{\frac{1}{9}} - \sqrt{0.36} - \sqrt{\frac{9}{49}} + \left(-\sqrt{32.49}\right) + \sqrt{3.24}$$

$$\frac{1}{3} - 0.6 - \frac{3}{7} - 5.7 + 1.8$$

$$\frac{1}{3} - \frac{3}{7} - 4.5$$

$$\frac{1}{3} \cdot \frac{70}{70} - \frac{3}{7} \cdot \frac{30}{30} - \frac{45}{10} \cdot \frac{21}{21}$$

$$\frac{70 - 90 - 945}{210}$$

$$-\frac{465}{210} = -\frac{193}{42} = -4\frac{25}{42}$$

71. Simplify:

$$3\sqrt{16} + 4\sqrt{25} - 8\sqrt{4}$$

$$3(4) + 4(5) - 8(2)$$

$$12 + 20 - 16$$

$$16$$

73. $\sqrt[3]{64} = 4$ $\sqrt[3]{125} = 5$ $\sqrt[3]{1000} = 10$

    $\sqrt[4]{16} = 2$ $\sqrt[4]{81} = 3$ $\sqrt[5]{32} = 2$

75. $4^2 + 5^2 = 16 + 25 = 41$

77. $2^2 + 3^2 + 4^2 = 4 + 9 + 16 = 29$

Chapter 6
True-False Concept Review

1. False – It is common to write the decimal 6.7 in words: six and seven tenths

2. True

3. True

4. False – A decimal point is the the separation of the ones place from the tenths place.

5. True

6. False – To the nearest ten, 74.49 rounds to 70.

7. True

8. False – the rounded value of a decimal is sometimes smaller or larger than the original decimal.

9. True

10. True

11. True

12. False – It is possible to subtract 9.4 from 6.0.

13. True

14. False – The quotient of 0.6 and 0.5 is 1.2/

15. False – Scientists are among a variety of groups who use scientific notation.

16. False – It is sometimes necessary to move the decimal point when dividing decimals.

17. False – It is not always necessary to round off when dividing decimals.

18. True

19. False – The procedures for dividing decimals and dividing fractions is different.

20. True

21. False – The quotient of two decimals may be smaller than either of the two decimals

22. False – There are two basic rules for rounding: "four-five" and "rounding by truncation."

23. False – The sum of 8.15 and 0.3 is 8.45

24. False – The product of two decimals may have more decimal places than either decimal (the product of 3.8 and 4.5 is 17.1); the product has the same number of places as either of the two factors; or the product may have no decimal places (the product of 3.2 and 0.625 is 2)

25. False – The smaller of two decimals may have the greater number of decimal places (0.2 < 8.555 is true)

26. True

27. False – There are four decimal places in the product of 7.61 and 4.63.

28. False – Some square roots are equal to or greater than the original number; such as $\sqrt{1} = 1$ or $\sqrt{0.04} = 0.2$

29. False – The Pythagorean Theorem can be used to find the third side of any right triangle.

30. True

Chapter 6
Review

1. seven hundred twenty-one thousandths = 0.721

3. eighteen and six hundred two ten-thousandths = 18.0602

5. 344.00082 = three hundred forty-four and eighty-two hundred-thousandths

7. $0.34 = \dfrac{34}{100} = \dfrac{17}{50}$

9. $0.875 = \dfrac{875}{1000} = \dfrac{7}{8}$

11. 0.2998   0.3409   0.3426   0.345

13. 0.0907   0.090733   0.0909   0.09099   0.0976

15. 83.25 > 83.035 is true

17. 0.0445 rounded to the nearest hundredth is 0.04

19. 2632.9378 rounded to the nearest thousandth is 2632.938

21. 6.994 + 12.536 + 4.64 + 0.8993 = 25.0693

23. 5.16 + 0.006824 + 11.20367 + 3.55546 + 9.9 = 29.825954

25. $-8.3t + 6.04t + (-0.55t) + 1.925t = -0.885t$

27. 39 − 17.0354 = 21.9646

29. $33v − 0.961v = 32.039v$

31. 0.06 + 0.01 + 0.07 + 0.02 = 0.16

33. 0.6 + 0.2 + 0.0 + 0.0 + 0.1 = 0.9

35. 0.1 − 0.8 − 0.6 + 0.3 = − −1

37. 4.2(0.484) = 2.0328

38. 0.64(22.8) = 14.592

39. 1.022(0.025) = 0.02555

41. 0.04(0.8) = 0.032

43. 0.05(0.3)(0.0004) = 0.000006

45. 1(0.9)(0.04) = 0.036

47. 0.0092(1000) = 9.2

49. $3.467 \div 10^{-6} = 3{,}467{,}000$

51. $40{,}000 = 4.0 \times 10^{4}$

53. $0.00000007 = 7.0 \times 10^{-8}$

55. $6.78 \times 10^{6} = 6{,}780{,}000$

57. $400.96 \div 32 = 12.53$

59. $5.63 \div 2.41 = 2.336$

61. $\dfrac{7}{16} = 0.4375$

63. $\dfrac{23}{64} = 0.359375$

65. $\dfrac{19}{29} = 0.6551 \approx 0.66$

67. $0.9 \div 3 = 0.3$

69. $4.9 \div 0.07 = 70$

71. $6.7\,\text{cm} \cdot \dfrac{1\,\text{km}}{10^{5}\,\text{cm}} = 0.000067\,\text{km}$

73. $\dfrac{4.7\,\text{oz}}{1} \cdot \dfrac{28.35\,\text{g}}{1\,\text{oz}} = 133.45\text{g}$

75. $\dfrac{34\,\text{mi}}{\text{hr}} \cdot \dfrac{1.609\,\text{km}}{\text{mi}} \cdot \dfrac{1000\,\text{m}}{\text{km}} \cdot \dfrac{\text{hr}}{60\,\text{min}} = \dfrac{911.77\,\text{m}}{\text{min}}$

77: Evaluate:
$$a(w + y)^{2}$$
$$0.06(43.5 + 1.2)^{2}$$
$$0.06(44.7)^{2}$$
$$0.06(1998.09)$$
$$119.8854$$

79. Evaluate:
$$P = 2\pi r$$
$$P = 2(3.14)(6.6)$$
$$P = 41.448$$

81. Solve:
$$2.25x + 3.9 = 4.125$$
$$2.25x = 0.225$$
$$x = 0.1$$

83. Solve:
$$1.3z + 1.466 = 2.22$$
$$1.3z = 0.104$$
$$z = 0.08$$

85. Solve:
$$4.1s + 51.6 + 0.7s = 150$$
$$4.8s + 51.6 = 150$$
$$4.8s = 98.4$$
$$s = 20.5$$

87. $\sqrt{900} = 30$

89. $\sqrt{236} \approx 15.36$

91. Solve:
$$a^2 + b^2 = c^2$$
$$a^2 + 16^2 = 34^2$$
$$a^2 + 256 = 1156$$
$$a^2 = 900$$
$$a = 30$$

93. Solve:
$$a^2 + b^2 = c^2$$
$$32^2 + b^2 = 43^2$$
$$1024 + b^2 = 1849$$
$$b^2 = 825$$
$$b = 28.722$$
$$b \approx 28.72$$

95. Solve:
$$a^2 + b^2 = c^2$$
$$500^2 + 375^2 = c^2$$
$$250,000 + 140,625 = c^2$$
$$390,625 = c^2$$
$$625 \text{ ft} = c$$

97. This is a right triangle.
$$a^2 + b^2 = c^2$$
$$11^2 + 26.4^2 = 28.6^2$$
$$121 + 696.96 = 817.96$$
$$817.96 = 817.96$$

Chapter 6
Test

1. $0.6523 \div 0.79 = 0.8256 \Rightarrow 0.826$

2. $\dfrac{12}{17} = 0.7058 \Rightarrow 0.71$

3. $0.067 \quad 0.06729 \quad 0.0673 \quad 0.06893$

4. $106.00408 =$ one hundred six and four hundred eight hundred-thousandths

5. $7.89 \times 3.45 = 27.2205$

6. Evaluate:
$$S = \pi r l$$
$$S = 3.14(2.4)(0.8)$$
$$S = 6.0288$$

7. 4.998 rounded to the nearest hundredth is 5.00

8. $0.024 = \dfrac{24}{1000} = \dfrac{3}{125}$

9. $12 - 6.7834 = 5.2166$

10. Solve:
$$0.05x - 0.008 = 0.082$$
$$0.05x = 0.09$$
$$x = 1.8$$

11. $0.0000071 = 7.1 \times 10^{-6}$

12. 1 is not a solution
$$3.84 + 0.45y - 4.5 \neq 4.24 - 1.45y$$
$$3.84 + 0.45(1) - 4.5 \neq 4.24 - 1.45(1)$$
$$3.84 + 0.45 - 4.5 \neq 4.24 - 1.45$$
$$-0.21 \neq 2.79$$

13. Solve:
$$11.3 - 2.4a = 9.16 + 3.6a - 11$$
$$22.3 - 2.4a = 9.16 + 3.6a$$
$$13.14 = 6a$$
$$2.19 = a$$

14. $8.773 - 3.4891 = 5.2839$

15. $\sqrt{1622} \approx 40.274 \approx 40.27$

16. $43820 = 4.382 \times 10^{4}$

17. $0.00567 < 0.0056099$ is false

18. $3.332 \div 5.3 = 0.62867 \Rightarrow 0.6287$

19. $7.26(-0.0085) = -0.06171$

20. $6.9 + 0.08 + 21.64 + 13.89 + 9.84 + 0.234 = 52.584$

21. $0.026 \div 100{,}000 = 0.00000026$

22. $-6.3a - 8.4 - 5.6a + 1.23 - 0.87a + 0.03 = -12.77a - 7.14$

23. $\dfrac{198.7 \text{ mi} + 203.5 \text{ mi} + 386.2 \text{ mi} + 187.6 \text{ mi}}{4} = \dfrac{976 \text{ mi}}{4} = 244 \text{ mi}$

24. $6.3 \times 10^{4} = 63{,}000$

25. $\dfrac{21}{3}(2.12) = 7(\$2.12) = \$14.84$

26. The missing side is:
$$a^2 + b^2 = c^2$$
$$320^2 + b^2 = 400^2$$
$$102{,}400 + b^2 = 160{,}000$$
$$b^2 = 57{,}600$$
$$b = 240$$

CHAPTER 1 - 6
CUMULATIVE REVIEW

1. $287,064 \div 36 = 7974$

3. $72,198 \times 302 = 21,803,796$

5. $963 > 936$ is true

7. $3x(4x + 3y - 8z) = 12x^2 + 9xy - 24xz$

9. Solve:
$$8a - 3 = 29$$
$$8a = 32$$
$$a = 4$$

11. 3 yd 2 ft + 1 yd 2 ft + 2 yd 1 ft = 6 yd 5 ft = 7 yd 2 ft

13. The average yield per acre on the five fields is:
$$\frac{100(65) + 150(70) + 85(55) + 125(83) + 62(90)}{522}$$
$$\frac{6500 + 10,500 + 4675 + 10,375 + 5580}{522}$$
$$\frac{37630}{522} = 72.08 \approx 72 \text{ bu}$$

15. $20(8) + \dfrac{1}{2}(10)(6) = 160 + 30 = 190 \text{ in}^2$

17. The second quarter has the highest point production.

19. Two fewer points are scored, on the average, in the first quarter than the fourth quarter $(30 - 28)$.

21. The average number of points per game is: $28 + 32 + 25 + 30 = 115$

23. $-67 - (-34) - 13 - (-72) - 53 = -67 + (+34) - 13 + (+72) - 53 = -133 + 106 = -27$

25. $15(-34 - [-54]) = 15(-34 + [+54]) = 15(20) = 300$

27. $\dfrac{24}{24+30} = \dfrac{24}{54} = \dfrac{4}{9}$

29. The first five multiples of 235 are: 235, 470, 705, 940, 1175

31. 1999 is a prime number

33. $\dfrac{36xy}{54x} = \dfrac{2y}{3}$

35. $\left(-\dfrac{5}{21}\right) \div \dfrac{3}{20} = \left(-\dfrac{5}{21}\right) \cdot \dfrac{20}{3} = -\dfrac{100}{63}$

37. $8\dfrac{3}{5} + 2\dfrac{3}{10} + 9\dfrac{3}{20} = 19 + \dfrac{3}{5} \cdot \dfrac{4}{4} + \dfrac{3}{10} \cdot \dfrac{2}{2} + \dfrac{3}{20} = 19 + \dfrac{12}{20} + \dfrac{6}{20} + \dfrac{3}{20} = 19\dfrac{21}{20} = 20\dfrac{1}{20}$

39. $45\dfrac{7}{15} - 21\dfrac{5}{6} = 45\dfrac{7}{15} \cdot \dfrac{2}{2} - 21\dfrac{5}{6} \cdot \dfrac{5}{5} = 45\dfrac{14}{30} - 21\dfrac{25}{30} = 44\dfrac{44}{30} - 21\dfrac{25}{30} = 23\dfrac{19}{30}$

41. The amount of cement needed is:

$$8\dfrac{3}{10} - 5\dfrac{3}{8}$$

$$8\dfrac{3}{10} \cdot \dfrac{4}{4} - 5\dfrac{3}{8} \cdot \dfrac{5}{5}$$

$$8\dfrac{12}{40} - 5\dfrac{15}{40}$$

$$7\dfrac{52}{40} - 5\dfrac{15}{40}$$

$$2\dfrac{37}{40} \text{ yd}$$

43. Solve:

$$3x - \dfrac{3}{4} = 18$$

$$4 \cdot 3x - 4 \cdot \dfrac{3}{4} = 4 \cdot 18$$

$$12x - 3 = 72$$

$$12x = 75$$

$$x = \dfrac{75}{12} = \dfrac{25}{4}$$

45. The width is:
$$P = 2l + 2w$$
$$17 = 2\left(6\frac{4}{9}\right) + 2w$$
$$17 = 2\left(\frac{58}{9}\right) + 2w$$
$$17 \cdot 9 = 2\left(\frac{58}{9} \cdot 9\right) + 2 \cdot 9w$$
$$153 = 116 + 18w$$
$$37 = 18w$$
$$2\frac{1}{18} \text{ ft} = w$$

47. Sixteen and twelve thousandths = 16.012

49. $0.48 = \dfrac{48}{100} = \dfrac{12}{25}$

51. 8.9049 rounded to the nearest hundredth is 8.90

53. $8.9053 - 6.9144 = 1.9909$

55. $1.0025(100,000) = 100,250$

57. $823,467 \div 1000 = 823.467$

59. $93,000,000 = 9.3 \times 10^7$

61. $2.8765 \div 0.55 = 5.23$

63. $\dfrac{17}{64} = 0.265625$

65. $V = \dfrac{4}{3}\pi r^3 = \dfrac{4}{3}(3.14)(0.25)^3 = \dfrac{4}{3}(3.14)(0.015625) = 0.0654$

67. $674 + 593 + 127 + 57 + 52 = 1503$

69. The budget for food, lodging, and car expenses is:
$$5 \text{ da}\left(\frac{\$72}{\text{da}}\right) + 6 \text{ da}\left(\frac{\$23}{\text{da}}\right) + 1350 \text{ mi} \cdot \frac{\text{gal}}{28 \text{ mi}} \cdot \frac{\$1.495}{\text{gal}}$$
$$\$360 + \$138 + \$72.08$$
$$\$570.08$$
The amount that Catherine has for entertainment is:
$$\$800 - \$570.08 = \$229.92$$

CHAPTER SEVEN
RATIO, PROPORTION, AND PERCENT

Section 7.1
Ratio and Rate

1. $\dfrac{7}{35} = \dfrac{1}{5}$

3. $\dfrac{12\,m}{10\,m} = \dfrac{6}{5}$

5. $\dfrac{40\,cents}{45\,cents} = \dfrac{8}{9}$

7. $\dfrac{1\,dime}{4\,nickels} = \dfrac{10\,cents}{20\,cents} = \dfrac{1}{2}$

9. $\dfrac{16\,in}{2\,ft} = \dfrac{16\,in}{24\,in} = \dfrac{2}{3}$

11. $\dfrac{200\,cm}{3\,km} = \dfrac{200\,cm}{300,000\,cm} = \dfrac{1}{1500}$

13. $\dfrac{8\,people}{11\,chairs}$

15. $\dfrac{110\,mi}{2\,hr} = \dfrac{55\,mi}{1\,hr}$

17. $\dfrac{63\,mi}{3\,gal} = \dfrac{21\,mi}{1\,gal}$

19. $\dfrac{88\,lb}{33\,ft} = \dfrac{8\,lb}{3\,ft}$

21. $\dfrac{18\,trees}{63\,ft} = \dfrac{2\,trees}{7\,ft}$

23. $\dfrac{38\,books}{95\,students} = \dfrac{2\,books}{5\,students}$

25. $\dfrac{765\,people}{27\,rooms} = \dfrac{85\,people}{3\,rooms}$

27. $\dfrac{345\,pies}{46\,sales} = \dfrac{15\,pies}{2\,sales}$

29. $\dfrac{50\,mi}{2\,hr} = \dfrac{25\,mi}{1\,hr}$

31. $\dfrac{36\,ft}{9\,sec} = \dfrac{4\,ft}{1\,sec}$

33. $\dfrac{132\,yd}{2\,billboards} = \dfrac{66\,yd}{1\,billboards}$

35. $\dfrac{90\,cents}{10\,lb} = \dfrac{9\,cents}{1\,lb}$

37. $\dfrac{825\,mi}{22\,gal} = \dfrac{37.5\,mi}{1\,gal}$

39. $\dfrac{1000\,ft}{12\,sec} \approx \dfrac{83.3\,ft}{1\,sec}$

41. $\dfrac{12,095\,lb}{45\,mi^2} \approx \dfrac{268.8\,lb}{1\,mi^2}$

43. $\dfrac{225\,\text{gal}}{14\,\text{min}} \approx \dfrac{16.1\,\text{gal}}{1\,\text{min}}$

45. a. The ratio of compact spaces to larger spaces is $\dfrac{18}{24} = \dfrac{3}{4}$

    b. The ratio of compact spaces to the total number of spaces is $\dfrac{18}{18+24} = \dfrac{18}{42} = \dfrac{3}{7}$

47. The rates of TV sets per house is not the same between the two sections of the country since $\dfrac{3500}{1000} = \dfrac{7}{2}$ and $\dfrac{500}{150} = \dfrac{10}{3}$

49. The ratio of the cost to the selling price is $\dfrac{\$175}{\$300} = \dfrac{7}{12}$

51. The ratio of the sale price to the regular price is $\dfrac{\$66.66}{\$99.99} = \dfrac{2}{3}$

53. The ratio of the number of people to the square miles is $\dfrac{22,450\,\text{people}}{230\,\text{mi}^2} \approx \dfrac{97.6\,\text{people}}{1\,\text{mi}^2}$

55. answers vary

57. a. The ratio of laundry use to toilet use is $\dfrac{35\,\text{gal}}{100\,\text{gal}} = \dfrac{17}{20}$

    b. The ratio of bathing/showering use to dish-washing use is $\dfrac{80\,\text{gal}}{15\,\text{gal}} = \dfrac{16}{3}$

59. The amount of lead needed to pollute 25 liters of drinking water is
0.05 mg/liter(25 liters) = 1.25 mg

61. The sixteen ounce can of pears is the better buy since:
$\dfrac{\$0.89}{14\,\text{oz}} \approx \dfrac{\$0.06357}{\text{oz}}$ and $\dfrac{\$1.00}{16\,\text{oz}} \approx \dfrac{\$0.0625}{\text{oz}}$

63. The five pound box of sugar is the better buy since:
$\dfrac{\$4.95}{5\,\text{lb}} \approx \dfrac{\$0.99}{\text{lb}}$ and $\dfrac{\$24.90}{25\,\text{lb}} \approx \dfrac{\$0.996}{\text{lb}}$

65. The unit price of a 12-oz can of orange juice is not the same as the unit price of a 16-oz can of orange juice since $\dfrac{\$1.09}{12\,oz} \approx \dfrac{\$0.0908}{oz}$ and $\dfrac{\$1.30}{16\,oz} \approx \dfrac{\$0.08125}{oz}$

67. answers vary

69. answers vary

71. Table about chicken sandwiches at various fast-food restaurants:

	Restaurant	Calories	Grams of Fat	Fat Calories	$\dfrac{\text{Fat Calories}}{\text{Total Calories}}$
a.	RB's Light	276	7	$7 \cdot 9 = 63$	$\dfrac{63}{276} = \dfrac{21}{92}$
b.	RB's Broiler	267	8	$8 \cdot 9 = 72$	$\dfrac{72}{267} = \dfrac{24}{89}$
c.	Hard B's	370	13	$13 \cdot 9 = 117$	$\dfrac{117}{370}$
d.	LJS's	130	4	$4 \cdot 9 = 36$	$\dfrac{36}{130} = \dfrac{18}{65}$
e.	The Major's	482	27	$27 \cdot 9 = 243$	$\dfrac{243}{482}$
f.	Mickey's	415	19	$19 \cdot 9 = 171$	$\dfrac{171}{415}$
g.	Tampico's	213	10	$10 \cdot 9 = 90$	$\dfrac{90}{213} = \dfrac{30}{71}$
h.	Winston's	290	7	$7 \cdot 9 = 63$	$\dfrac{63}{290}$

73. answers vary

75. Solve:

$$\frac{x}{2} - \frac{2}{3} = \frac{3}{4}$$
$$12 \cdot \frac{x}{2} - 12 \cdot \frac{2}{3} = 12 \cdot \frac{3}{4}$$
$$6x - 8 = 9$$
$$6x = 17$$
$$x = \frac{17}{6}$$

77. Solve:

$$\frac{y}{10} + \frac{6}{25} = -\frac{3}{2}$$
$$50 \cdot \frac{y}{10} + 50 \cdot \frac{6}{25} = -\frac{3}{2} \cdot 50$$
$$5y + 12 = -75$$
$$5y = -87$$
$$y = -\frac{87}{5}$$

Section 7.2
Solving Proportions

1. The proportion is true.
$$\frac{6}{3} = \frac{16}{8}$$
$$3 \cdot 16 = 6 \cdot 8$$
$$48 = 48$$

3. The proportion is true.
$$\frac{6}{8} = \frac{9}{12}$$
$$8 \cdot 9 = 6 \cdot 12$$
$$72 = 72$$

5. The proportion is false.
$$\frac{3}{2} \neq \frac{9}{4}$$
$$3 \cdot 4 \neq 2 \cdot 9$$
$$12 \neq 18$$

7. The proportion is true.
$$\frac{18}{15} = \frac{12}{10}$$
$$18 \cdot 10 = 15 \cdot 12$$
$$180 = 180$$

9. The proportion is false.
$$\frac{35}{22} \neq \frac{30}{20}$$
$$35 \cdot 20 \neq 22 \cdot 30$$
$$700 \neq 660$$

11. The proportion is false.
$$\frac{27}{45} \neq \frac{30}{60}$$
$$27 \cdot 60 \neq 45 \cdot 30$$
$$1620 \neq 1350$$

13. The proportion is true.
$$\frac{2.8125}{3} = \frac{15}{16}$$
$$3 \cdot 15 = 16 \cdot 2.8125$$
$$45 = 45$$

15. Solve:
$$\frac{1}{2} = \frac{a}{18}$$
$$2a = 1 \cdot 18$$
$$2a = 18$$
$$a = 9$$

17. Solve:
$$\frac{2}{6} = \frac{c}{18}$$
$$6c = 2 \cdot 18$$
$$6c = 36$$
$$c = 6$$

19. Solve:
$$\frac{2}{a} = \frac{5}{10}$$
$$5a = 2 \cdot 10$$
$$5a = 20$$
$$a = 4$$

21. Solve:

$$\frac{14}{28} = \frac{5}{c}$$

$$14c = 28 \cdot 5$$

$$14c = 140$$

$$c = 10$$

23. Solve:

$$\frac{2}{3} = \frac{12}{x}$$

$$2x = 3 \cdot 12$$

$$2x = 36$$

$$x = 18$$

25. Solve:

$$\frac{w}{7} = \frac{2}{28}$$

$$28w = 7 \cdot 2$$

$$28w = 14$$

$$w = \frac{1}{2}$$

27. Solve:

$$\frac{x}{7} = \frac{3}{2}$$

$$2x = 7 \cdot 3$$

$$2x = 21$$

$$x = 10.5$$

29. Solve:

$$\frac{2}{z} = \frac{5}{11}$$

$$5z = 2 \cdot 11$$

$$5z = 22$$

$$z = 4.4$$

31. Solve:

$$\frac{16}{24} = \frac{y}{16}$$

$$24y = 16 \cdot 16$$

$$24y = 256$$

$$y = \frac{256}{24} = \frac{32}{3}$$

33. Solve:

$$\frac{15}{16} = \frac{12}{a}$$

$$15a = 16 \cdot 12$$

$$15a = 192$$

$$a = \frac{192}{15} = 12.8$$

35. Solve:

$$\frac{0.1}{c} = \frac{0.2}{1.2}$$

$$0.2c = 1.2(0.1)$$

$$0.2c = 0.12$$

$$c = \frac{0.12}{0.2} = 0.6$$

37. Solve:

$$\frac{\frac{3}{5}}{b} = \frac{8}{5}$$

$$8b = \frac{3}{5}(5)$$

$$8b = 3$$

$$b = \frac{3}{8}$$

39. Solve:

$$\frac{0.9}{4.5} = \frac{0.05}{x}$$

$$0.9x = 4.5(0.05)$$

$$0.9x = 0.225$$

$$x = 0.25$$

41. Solve:

$$\frac{7}{42} = \frac{w}{3}$$
$$42w = 7 \cdot 3$$
$$42w = 21$$
$$w = 0.5$$

43. Solve:

$$\frac{y}{3} = \frac{9}{\frac{1}{8}}$$
$$\frac{1}{8} y = 3 \cdot 9$$
$$\frac{1}{8} y = 27$$
$$y = 27 \cdot \frac{8}{1} = 216$$

45. Solve:

$$\frac{t}{24} = \frac{3\frac{1}{2}}{10\frac{1}{2}}$$
$$\frac{21}{2} t = 24 \cdot \frac{7}{2}$$
$$\frac{21}{2} t = 84$$
$$t = \frac{84}{1} \cdot \frac{2}{21} = 8$$

47. Solve:

$$\frac{3}{11} = \frac{w}{5}$$
$$11w = 3 \cdot 5$$
$$11w = 15$$
$$w \approx 1.36 \Rightarrow 1.4$$

49. Solve:

$$\frac{9}{35} = \frac{14}{y}$$
$$9y = 35 \cdot 14$$
$$9y = 490$$
$$y \approx 54.4$$

51. Solve:

$$\frac{1.5}{5.5} = \frac{a}{0.8}$$
$$5.5a = 1.5 \cdot 0.8$$
$$5.5a = 1.2$$
$$a \approx 0.22$$

53. Solve:

$$\frac{\frac{3}{7}}{c} = \frac{9}{20}$$

$$9c = \frac{3}{7} \cdot 20$$

$$9c = \frac{60}{7}$$

$$c = \frac{60}{7} \cdot \frac{1}{9}$$

$$c = \frac{60}{63} \approx 0.95$$

55. Solve:

$$\frac{?}{70} = \frac{1}{7}$$

$$7 \cdot ? = 70 \cdot 1$$

$$7 \cdot ? = 70$$

$$? = 10$$

57. The correct statement is:

$$\frac{2}{5} = \frac{x}{19}$$

$$5 \cdot x = 2 \cdot 19$$

59. 4.5 lbs of bananas cost:

$$\frac{2.4}{\$1.66} = \frac{4.5}{x}$$

$$2.4x = 4.5(\$1.66)$$

$$2.4x = \$7.47$$

$$x \approx \$3.11$$

61. The maximum price is:

$$\frac{18}{\$3.65} = \frac{25}{x}$$

$$18x = 25(\$3.65)$$

$$18x = \$91.25$$

$$x = \$5.069 \Rightarrow \$5.07$$

63. answers vary

65. Solve:

$$\frac{9+3}{9+6} = \frac{8}{a}$$

$$\frac{12}{15} = \frac{8}{a}$$

$$12a = 8 \cdot 15$$

$$12a = 120$$

$$a = 10$$

67. Solve:

$$\frac{7}{w} = \frac{18.92}{23.81}$$

$$18.92w = 7 \cdot 23.81$$

$$18.92w = 166.67$$

$$w \approx 8.809$$

69. answers vary

71. Solve:

$$\frac{x}{9} - \frac{4}{18} = \frac{1}{9}$$

$$18 \cdot \frac{x}{9} - 18 \cdot \frac{4}{18} = 18 \cdot \frac{1}{9}$$

$$2x - 4 = 2$$

$$2x = 6$$

$$x = 3$$

73. $4.835(10,000) = 48,350$

Section 7.3
Applications of Proportions

1. Box (a) = 6 in.

3. Box (c) = 15 in.

5. $\dfrac{6}{4} = \dfrac{15}{x}$

7. Box (1) = 30 ft

9. Box (3) = x

11. $\dfrac{30}{18} = \dfrac{x}{48}$

13. Box (5) = 8 hr

15. Box (7) = x

17. $\dfrac{8}{48} = \dfrac{x}{288}$

$$48x = 2305$$

$$x = 48 \text{ hr}$$

19. Box (e) = x

21. $\dfrac{3}{65} = \dfrac{x}{910}$

23. Four additional teachers will need to be assigned $(42 - 38 = 4)$

25. $\dfrac{1.5}{30} = \dfrac{14}{x}$

27. The number of rows that must be knitted is:

$$\frac{8}{1} = \frac{x}{12.5}$$

$$x = 8(12.5)$$

$$x = 100 \text{ rows}$$

29. The amount of fertilizer needed is:

$$\frac{16}{1500} = \frac{x}{2500}$$

$$1500x = 16(2500)$$

$$1500x = 40,000$$

$$x = 26\frac{2}{3} \text{ lb}$$

31. The chart is:

	Case I	Case II
Games Won	12	x
Games Played	15	30

33. The number of winning games is:

$$15x = 360$$

$$x = 24 \text{ games}$$

35. The number of weeks is:

$$125 \div 40 = 3\frac{1}{8}$$

$$4 \text{ weeks}$$

37. Hazel is paid:

$$\frac{1200}{60} = \frac{2900}{x}$$

$$1200x = 60 \cdot 2900$$

$$1200x = 174,000$$

$$x = \$145$$

39. The price is:

$$\frac{44}{4.84} = \frac{20}{x}$$

$$44x = 20 \cdot 4.84$$

$$44x = 96.8$$

$$x = \$2.20$$

41. The cost is:

$$\frac{25}{\$23.70} = \frac{10}{x}$$

$$25x = 10 \cdot \$23.70$$

$$25x = \$237.00$$

$$x = \$9.48$$

43. The number of gallons is:

$$\frac{1.5}{1} = \frac{9}{x}$$

$$1.5x = 9$$

$$x = 6 \text{ gal}$$

45. The cost to carpet the second room is:

$$\frac{33}{\$526.35} = \frac{22}{x}$$
$$33x = 22 \cdot \$526.35$$
$$33x = \$11{,}579.70$$
$$x = \$350.90$$

47. The number of cc's Ida should use is:

$$\frac{20\text{ mg}}{1\text{ cc}} = \frac{8\text{ mg}}{x}$$
$$20x = 1(8)$$
$$20x = 8$$
$$x = \frac{8}{20} = \frac{4}{10} = 0.4\text{ cc}$$

49. The cost of 1 lb of gumdrops is:

$$\frac{1.5}{\$0.45} = \frac{16}{x}$$
$$1.5x = 16(\$0.45)$$
$$1.5x = 720$$
$$x = 480\text{ cents} = \$4.80$$

51. The pounds of cashews needed is:

$$\frac{3}{7} = \frac{x}{40 - x}$$
$$7x = 3(40 - x)$$
$$7x = 120 - 3x$$
$$10x = 120$$
$$x = 12\text{ lb cashews}$$
The pounds of peanuts needed is
$$40 - 12 = 28\text{ lb}$$

53. The quarts of blue paint needed is:

$$\frac{3}{4} = \frac{x}{98 - x}$$
$$4x = 3(98 - x)$$
$$4x = 294 - 3x$$
$$7x = 294$$
$$x = 42\text{ qt}$$

55. The pounds of ground round is:

$$\frac{10}{13} = \frac{x}{91}$$
$$13x = 10 \cdot 91$$
$$13x = 910$$
$$x = 70\text{ lb}$$

57. The cost in pounds is:

$$\frac{1}{0.65} = \frac{2300}{x}$$
$$1x = 0.65 \cdot 2300$$
$$x = 1495\text{ pounds}$$

59. The comparable price is:

$$\frac{\$35.85}{36} = \frac{x}{60}$$
$$36x = 60(\$35.85)$$
$$36x = \$2151$$
$$x = \$59.75$$

61. The greatest amount of ozone is:

$$\frac{1}{235} = \frac{12}{x}$$
$$1x = 12 \cdot 235$$
$$x = 2820\text{ mg}$$

63. The price of one batch in the warehouse store is:

$$\frac{4}{4.79} = \frac{1}{x}$$

$$4x = 4.79$$

$$x = \$1.197$$

$$x \approx \$1.20$$

The grocery store should reduce the price of each box by $0.09 = \$1.29 - \$1.20$.

65. answers vary

67. The increase of live California condors from 1982 to 1992 is $73 - 25 = 48$.
The increase of live California condors from 1992 to 2017 will be:

$$\frac{48 \text{ condors}}{10 \text{ yr}} = \frac{x \text{ condors}}{25 \text{ yr}}$$

$$10x = 1200$$

$$x = 120 \text{ condors}$$

The number of condors alive in 2017 will be $(120 + 73)$ condors $= 193$ **condors**.

69. The cost for the Santini family was:

$$\frac{8 \text{ da}}{19 \text{ da}} = \frac{x}{\$1905}$$

$$19x = 15,240$$

$$x = \$802$$

The cost for the Nguyen family was:

$$\frac{11 \text{ da}}{19 \text{ da}} = \frac{x}{\$1905}$$

$$19x = 20,955$$

$$x = \$1103$$

71. answers vary

73. 37.4145 rounded to the nearest hundredth is 37.41

75. $-0.066 > -0.15$

Section 7.4
Percents, Decimals, and Fractions

1. $\dfrac{19}{100} = 19\%$

3. $\dfrac{1}{2} = 50\%$

5. $\dfrac{7}{10} = 70\%$

7. $\dfrac{9}{25} = 36\%$

9. $\dfrac{2}{5} = 40\%$

11. $\dfrac{18}{25} = 72\%$

13. $\dfrac{3}{8} = 37\dfrac{1}{2}\%$

15. $0.37 = 37\%$

17. $2.34 = 234\%$

19. $1.4 = 140\%$

21. $2 = 200\%$

23. $0.052 = 5.2\%$

25. $0.1025 = 10.25\%$

27. $0.001 = 0.1\%$

29. $0.0001 = 0.01\%$

31. $\dfrac{2}{3} = 0.667 = 66.7\%$

33. $\dfrac{5}{12} = 0.417 = 41.7\%$

35. $\dfrac{4}{7} = 0.571 = 57.1\%$

37. $17\% = 0.17$

39. $2.35\% = 0.0235$

41. $315\% = 3.15$

43. $0.12\% = 0.0012$

45. $0.25\% = 0.0025$

47. $1\% = 0.01$

49. $37\% = \dfrac{37}{100}$

51. $10\% = \dfrac{10}{100} = \dfrac{1}{10}$

53. $4.756\% = 0.04756$

55. $3\dfrac{4}{5}\% = 3.8\% = 0.038$

57. $46\dfrac{3}{8}\% = 46.375\% = 0.46375$

59. $33\dfrac{1}{3}\% = \dfrac{100}{3} \cdot \dfrac{1}{100} = \dfrac{1}{3}$

61. $15\dfrac{2}{3}\% = \dfrac{47}{3} \cdot \dfrac{1}{100} = \dfrac{47}{300}$

63. $0.02\% = 0.0002 = \dfrac{2}{10,000} = \dfrac{1}{5000}$

65. $\dfrac{28}{40} = 70\%$

67. $0.022 = 2.2\%$

69. $\dfrac{18}{100} = 18\%$

71. $0.375 = 37.5\%$

73. $7.5\% = 0.075$

75. $1.87 = 187\%$

77. $0.3\% = 0.003$

79. $0.6\% = 0.006$

81. $3 = 300\%$

83. $\dfrac{7\cdot 9}{290}=\dfrac{63}{290}=21.7\% \Rightarrow 22\%$ **85.** $42\%\ \dfrac{42}{100}=\dfrac{21}{50}$

87. answers vary   **89.** answers vary

91. $24\dfrac{7}{8}\% = 24.875\% = 0.24875 = \dfrac{24{,}875}{100{,}000} = \dfrac{199}{800}$

93. answers vary

95. $7.653t^2 + 89.603t + \left(-1.863t^2\right) + \left(-27.536t\right) = 5.79t^2 + 62.067t$

97. Solve:
$$1.25y + 3.07 = 7.57$$
$$1.25y = 4.5$$
$$y = 3.6$$

Section 7.5
Percents, Decimals, and Fractions

1. Solve:
$$9 = 50\%B$$
$$9 = 0.5B$$
$$\dfrac{9}{0.5} = \dfrac{0.5B}{0.5}$$
$$18 = B$$

3. Solve:
$$3 = R(1)$$
$$3 = R$$
$$300\% = R$$

5. Solve:
$$R(60) = 30$$
$$\dfrac{60R}{60} = \dfrac{30}{60}$$
$$R = \dfrac{1}{2}$$
$$R = 50\%$$

7. Solve:
$$70\%B = 28$$
$$0.7B = 28$$
$$\dfrac{0.7B}{0.7} = \dfrac{28}{0.7}$$
$$B = 40$$

9. Solve:
$$80\%(45) = P$$
$$0.8(45) = P$$
$$36 = P$$

11. Solve:
$$R(80) = 64$$
$$\dfrac{80R}{80} = \dfrac{64}{80}$$
$$R = \dfrac{8}{10} = \dfrac{4}{5}$$
$$R = 80\%$$

13. Solve:

$$19\% B = 19$$
$$0.19B = 19$$
$$\frac{0.19B}{0.19} = \frac{19}{0.19}$$
$$B = 100$$

15. Solve:

$$0.5\%(200) = P$$
$$0.005(200) = P$$
$$1 = P$$

17. Solve:

$$70\%(40) = P$$
$$0.7(40) = P$$
$$28 = P$$

19. Solve:

$$140\% B = 56$$
$$1.4B = 56$$
$$\frac{1.4B}{1.4} = \frac{56}{1.4}$$
$$B = 40$$

21. Solve:

$$9.5\%(40) = P$$
$$0.095(40) = P$$
$$3.8 = P$$

23. Solve:

$$R(40) = 0.4$$
$$\frac{40R}{40} = \frac{0.4}{40}$$
$$R = 0.01$$
$$R = 1\%$$

25. Solve:

$$65\% B = 78$$
$$0.65B = 78$$
$$\frac{0.65B}{0.65} = \frac{78}{0.65}$$
$$B = 120$$

27. Solve:

$$R(1000) = 10$$
$$\frac{1000R}{1000} = \frac{10}{1000}$$
$$R = 0.01$$
$$R = 1\%$$

29. Solve:

$$48\%(45) = P$$
$$0.48(45) = P$$
$$21.6 = P$$

31. Solve:

$$37\% B = 222$$
$$0.37B = 222$$
$$\frac{0.37B}{0.37} = \frac{222}{0.37}$$
$$B = 600$$

33. Solve:

$$145\% B = 66.7$$
$$145B = 66.7$$
$$\frac{1.45B}{1.45} = \frac{66.7}{1.45}$$
$$B = 46$$

35. Solve:

$$R(125) = 66$$
$$\frac{125R}{125} = \frac{66}{125}$$
$$R = 0.528$$
$$R = 52.8\%$$

37. Solve:

$$14.4\%(75) = P$$
$$0.144(75) = P$$
$$10.8 = P$$

39. Solve:

$$3.6\%(0.8) = P$$
$$0.036(0.8) = P$$
$$0.0288 = P$$

41. Solve:

$$\frac{1}{2}\%(0.4) = P$$
$$0.5\%(0.4) = P$$
$$0.005(0.4) = P$$
$$0.002 = P$$

43. Solve:

$$11\frac{1}{9}\%(1548) = P$$

$$\frac{100}{9}\left(\frac{1}{100}\right)(1548) = P$$

$$\frac{1548}{9} = P$$

$$172 = P$$

45. Solve:

$$16\frac{2}{3}\%(4002) = P$$

$$\frac{50}{3}\left(\frac{1}{100}\right)(4002) = P$$

$$\frac{4002}{6} = P$$

$$667 = P$$

47. Solve:

$$1.25\%(1250) = P$$

$$0.0125(1250) = P$$

$$15.62 = P$$

$$15.6 \approx P$$

49. Solve:

$$16\%B = 9.3$$

$$0.16B = 9.3$$

$$\frac{0.16B}{0.16} = \frac{9.3}{0.16}$$

$$B \approx 58.1$$

51. Solve:

$$R(75) = 8$$

$$\frac{75R}{75} = \frac{8}{75}$$

$$R \approx 0.107$$

$$R \approx 10.7\%$$

53. Solve:

$$R(28) = 38$$

$$\frac{28R}{28} = \frac{38}{28}$$

$$R \approx 1.36$$

$$R \approx 136\%$$

55. Solve:

$$5\frac{1}{3}\%\left(6\frac{1}{2}\right) = P$$

$$\frac{16}{3}\%\left(6\frac{1}{2}\right) = P$$

$$\frac{16}{3}\left(\frac{1}{100}\right)\left(\frac{13}{2}\right) = P$$

$$\frac{104}{300} = P$$

$$\frac{26}{75} = P$$

57. answers vary

59. Solve:

$$\frac{1}{2}\%\left(33\frac{1}{3}\right) = P$$

$$\frac{1}{2}\left(\frac{1}{100}\right)\left(\frac{100}{3}\right) = P$$

$$\frac{1}{6} = P$$

61. Solve:

$$\frac{9}{10}\%B = 2\frac{1}{2}$$

$$\frac{9}{10} \cdot \frac{1}{100}B = \frac{5}{2}$$

$$\frac{9}{1000}B = \frac{5}{2}$$

$$\frac{9}{1000} \cdot \frac{1000}{9}B = \frac{5}{2} \cdot \frac{1000}{9}$$

$$B = \frac{2500}{9}$$

63. group activity

65. $27.82(-3.096) = -86.13072$

67. $0.5792 \div 0.016 = 36.2$

69. Solve:

$$16t - 18.4t = 5.76$$

$$-2.4t = 5.76$$

$$\frac{-2.4t}{-2.4} = \frac{5.76}{-2.4}$$

$$t = -2.4$$

Section 7.6
Applications of Percents

1. The tax is:

$$t = 4\%(\$129.95)$$

$$t = 0.04(\$129.95)$$

$$t = \$5.20$$

3. Joan's yearly salary is:

$$14\%s = \$2193.10$$

$$\frac{0.14s}{0.14} = \frac{\$2193.10}{0.14}$$

$$s = \$15,665$$

5. Maria's total earnings are:
   $$e = 40(\$7.82) + 7(1.5)(\$7.82)$$
   $$e = \$312.80 + \$82.11$$
   $$e = \$394.91$$

7. John's % score is:
   $$\frac{25}{30} = \frac{5}{6} = 0.833 = 83.3\% \Rightarrow 83\%$$

9. The % of the tip is:
   $$\frac{\$5}{\$33.45} = 0.15 = 15\%$$

11. The no. of men younger than 40 is:
    $$68\%(7002) = ,68(7002) = 4761 \text{ men}$$

13. The % savings is:
    $$\frac{\$20}{\$72.99} = 0.274 = 27.4\%$$

15. The number of physical therapists predicted by 2005 is:
    $$188\%(90,000) = 1.88(90,000) = 169,200 \Rightarrow 169,000$$

17. The amount of Merle's account for August is:
    $$\$2.80 = 1.75\%X$$
    $$\$2.80 = 0.0175X$$
    $$\frac{\$2.80}{0.0175} = \frac{0.0175X}{0.0175}$$
    $$\$160 = X$$

19. answers vary

21. The markup is: $\dfrac{\$5.04}{\$16.80} = 0.3 = 30\%$

23. The markup is: $28\%(\$39.99) = 0.28(\$39.99) = \$11.20$

25. The sale price of the TV is:
    $$(100\% - 27\%)(\$335.95) = 73\%(\$335.95) = 0.73(\$335.95) = \$245.24$$
    The TV in this problem is the better buy by: $\$247.46 - \$245.24 = \$2.22$

27. The commission is: $12\%(\$4725) = 0.12(\$4725) = \$567$

29. The percent tax rate is: $\dfrac{\$2238.30}{\$82900} = 0.027 = 2.7\%$

31. The percent off is: $\dfrac{\$59.99 - \$47.99}{\$59.99} = \dfrac{\$12}{\$59.99} = 0.20 = 20\%$

33. The price of the blazer is:
$$60\%[80\%(\$139.95)] = .6[.8(\$139.95)] = .6[\$111.96] = \$67.18$$
The percent saving over the original price is:
$$\frac{\$139.95 - \$67.18}{\$139.95} = \frac{\$72.77}{\$139.95} = 0.52 = 52\%$$

35. After the two discounts are applied, the price is:
$$75\%(\$279.95) - 10\%(279.95) = 0.75(\$279.95) - 0.1(279.95) =$$
$$\$209.96 - \$28 = \$181.97$$

37. The amount saved on the knit shirt is:
$$70\%[80\%(\$14.99)] = 0.7[0.8(\$14.99)] = 0.7[\$11.99] = \$8.39$$
$$\$14.99 - \$8.39 = \$6.60$$

39. The number of tons recycled is: $15\%(11 \text{ tons}) = 0.15(11 \text{ tons}) = 1.65 \text{ tons}$

41. The Hispanic group has the smallest number of children in head Start.

43. The greatest expenditure of funds occurs with salaries.

45. The amount spent on development is:
$$20\%(\$2,500,000) = 0.2(\$2,500,000) = \$500,000$$

47. The probability of having three girls is $\dfrac{1}{8} = 0.125 = 12.5\%$

The probability of having three boys is $\dfrac{1}{8} = 0.125 = 12.5\%$

The probability of having two girls and one boy is $\dfrac{3}{8} = 0.375 = 37.5\%$

The probability of having one girl and two boys is $\dfrac{3}{8} = 0.375 = 37.5\%$

The circle graph to illustrate the probabilities of girl/boy combinations in 3 births is:

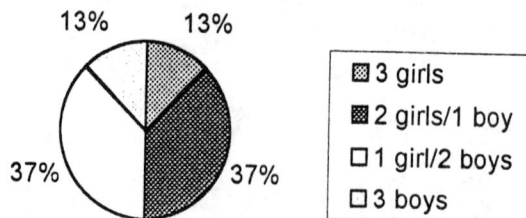

12.50%
13%   13%
37%   37%

☒ 3 girls
☒ 2 girls/1 boy
☐ 1 girl/2 boys
☐ 3 boys

49. The graph that illustrates the major causes of death worldwide is:

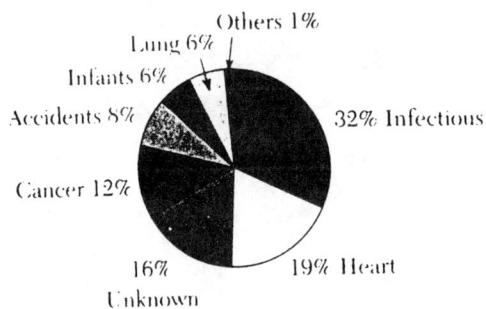

51. answers vary

53. The selling price is:
$$\$58 + 30\%x = x$$
$$\$58 + 0.3x = x$$
$$\$58 = 0.7x$$
$$\$82.86 = x$$

55. The percent of savings of the diesel compared with the gasoline engine is:
$$\frac{\$4600 - \$2650}{\$4600} = \frac{\$1950}{\$4600} = 0.42 = 42\%$$

57. group activity

59. Solve:
$$9.54 + 4t - 3.6t = 3.58$$
$$9.54 + 0.4t = 3.58$$
$$0.4t = -5.96$$
$$t = -14.9$$

61. Solve
$$0.34y + 0.55 + 0.76y - 0.3y = -0.32$$
$$0.8y + 0.55 = -0.32$$
$$0.8y = -0.87$$
$$y = -1.0875$$

Chapter 7
True-False Concept Review

1. True

2. True

3. False – $\dfrac{7 \text{ miles}}{1 \text{ gallon}} = \dfrac{14 \text{ miles}}{2 \text{ gallon}}$

4. True

5. False – If $\dfrac{7}{3} = \dfrac{t}{4}$, then 3t = 28 and $t = \dfrac{28}{3}$

6. True

7. False – All types of fractions can be written as percents by multiplying them by 100 and then inserting a % sign.

8. True

9. False – $7\dfrac{1}{2} = 7.5 = 750\%$

10. False – 0.009%=0.00009

11. False – If 0.3% of B is 84, then B = 28,000
$$0.3\%B = 84$$
$$0.003B = 84$$
$$B = 28,000$$

12. False – If R% of 12 is 24, then R = 200
$$R\%(12) = 24$$
$$\dfrac{R\%(12)}{12} = \dfrac{24}{12}$$
$$R\% = 200\%$$

13. True

14. False – Two consecutive decreases of 15% is the same as one decrease of $27\dfrac{3}{4}\%$.

15. False – If a salary is raised 10% and then subsequently lowered 10%, the ending salary is not the same as the beginning salary. It is 1% lower.

16. True

17. True

18. True

19. False – Ratios do not always compare like units.

20. True

21. False – $\dfrac{1}{2}\% = 0.5\% = 0.005$

22. True

23. True

24. True

25. False – To change a fraction to a percent, we must build a proportion: $\dfrac{a}{b} = \dfrac{r}{100}$

Chapter 7
Review

1. $\dfrac{63\,\text{ft}}{18\,\text{ft}} = \dfrac{7}{2}$

3. $\dfrac{2\,\text{dimes}}{\$1} = \dfrac{4\,\text{nickles}}{20\,\text{nickels}} = \dfrac{1}{5}$

5. $\dfrac{18\,\text{oz}}{2\,\text{lb}} = \dfrac{18\,\text{oz}}{32\,\text{oz}} = \dfrac{9}{16}$

7. $\dfrac{1898\,\text{cents}}{6\,\text{boxes}} = \dfrac{949\,\text{cents}}{3\,\text{boxes}}$

9. $\dfrac{234 \text{ VCR's}}{36 \text{ CD players}} = \dfrac{13 \text{ VCR's}}{2 \text{ CD players}}$

11. $\dfrac{\$12.95}{14 \text{ lb}} = 0.925 \text{ \$/lb}$

13. $\dfrac{44{,}896 \text{ people}}{460 \text{ sq. mi}} = 97.6 \text{ people/sq mi}$

15. $\dfrac{31{,}535 \text{ nautical mi}}{1802 \text{ gal}} = 17.5 \text{ n.mi./gal}$

17. $\dfrac{21}{36} = \dfrac{105}{70}$ is a false proportion

$21(170) \neq 36(105)$

$3570 \neq 3780$

19. $\dfrac{9.6}{5.7} = \dfrac{6.72}{4.05}$ is a false proportion

$9.6(4.05) \neq 5.7(6.72)$

$38.88 \neq 38.304$

21. Solve:

$\dfrac{x}{8} = \dfrac{13}{4}$

$4x = 8(13)$

$4x = 104$

$x = 26$

23. Solve:

$\dfrac{6.8}{1.5} = \dfrac{x}{4.5}$

$1.5x = 6.8(4.5)$

$1.5x = 30.6$

$x = 20.4$

25. Solve:

$\dfrac{0.024}{14} = \dfrac{x}{7}$

$14x = 7(0.024)$

$14x = 0.168$

$x = 0.012$

27. The profit is:

$\dfrac{110}{5} = \dfrac{x}{460}$

$5x = 110(460)$

$5x = 50{,}600$

$x = 10{,}120$

29. The number of miles is:

$\dfrac{1.5 \text{ in}}{100 \text{ mi}} = \dfrac{4 \text{ in}}{x}$

$1.5x = 4(100)$

$1.5x = 400$

$x = 266.\overline{6} \text{ mi}$

$x \approx 267 \text{ mi}$

31. $\dfrac{1}{8} = 0.125 = 12.5\%$

33. $\dfrac{15}{16} = 0.9375 = 93.75\%$

35. $\dfrac{101}{85} = 1.188 = 1188.8\%$

37. $0.0893 = 8.93\%$

39. $0.0009 = 0.09\%$

41. $0.375\% = 0.00375$

43. $134.9\% = 1.349$

45. $0.08642\% = 0.0008642$

47. Solve:

$$72 = 160R$$

$$\frac{72}{160} = R$$

$$0.45 = R$$

$$45\% = R$$

49. Solve:

$$73.4\%(123.7) = P$$

$$0.734(123.7) = P$$

$$90.8 = P$$

51. The % of readings above 225 is:

$$43 = 210R$$

$$\frac{43}{210} = R$$

$$0.205 = R$$

$$20.5\% = R$$

53. The former population was:

$$112\%B = 35{,}280$$

$$1.12B = 35{,}280$$

$$\frac{1.12B}{1.12} = \frac{35{,}280}{1.12}$$

$$B = 31{,}500$$

55. Hazel's monthly income is:

$$23\%B = \$350$$

$$0.23B = \$350$$

$$\frac{0.23B}{0.23} = \frac{\$350}{0.23}$$

$$B = \$1522$$

57. The commission is:

$$11\%(\$45{,}600) = P$$

$$0.11(\$45{,}600) = P$$

$$\$5016 = P$$

59. The balance of Leslie's account is:

$$1.25\%B = \$6.10$$

$$0.0125B = \$6.10$$

$$\frac{0.0125B}{0.125} = \frac{\$6.10}{0.125}$$

$$B = \$488$$

61. The Andrews spent 20% of their income on cars.

63. The Andrews spend more money on automobile expenses (by 5%).

65. The Andrews spend 50% on cars (20%) + rent (30%).

Chapter 7
Test

1. $\dfrac{15}{75} = \dfrac{1}{5}$

2. The proportion $\dfrac{16}{34} = \dfrac{24}{51}$ is true.

$$16(51) = 34(24)$$
$$816 = 816$$

3. $1.6 = 160\%$

4. 4. $\dfrac{1}{5} = \dfrac{1}{5} \cdot \dfrac{20}{20} = \dfrac{20}{100} = 20\%$

5. $78\% = 0.78$

6. $\dfrac{3\text{ qt}}{1\text{ gal}} = \dfrac{3\text{ qt}}{4\text{ qt}} = \dfrac{3}{4}$

7. $32.5\% = 0.325 = \dfrac{325}{1000} = \dfrac{13}{40}$

8. Solve:

$$\dfrac{2.1}{9} = \dfrac{0.56}{w}$$
$$2.1w = 9(0.56)$$
$$2.1w = 5.04$$
$$\dfrac{2.1w}{2.1} = \dfrac{5.04}{2.1}$$
$$w = 2.4$$

9. The number is:

$$62\%B = 58.9$$
$$0.62B = 58.9$$
$$\dfrac{0.62B}{0.62} = \dfrac{58.9}{0.62}$$
$$B = 95$$

10. $0.003 = 0.3\%$

11. $\dfrac{10 \text{ lb}}{35 \text{ lb}} = \dfrac{2}{7}$

12. $0.13\% = 0.0013$

13. The number is:
$$11.5\%(212) = P$$
$$0.115(212) = P$$
$$24.38 = P$$

14. Solve:
$$\dfrac{9}{24} = \dfrac{x}{28}$$
$$24x = 9(28)$$
$$24x = 252$$
$$\dfrac{24x}{24} = \dfrac{252}{24}$$
$$x = 10.5$$

15. The percent is:
$$\dfrac{7}{23}$$
$$0.3043$$
$$30.43\%$$
$$30.4\%$$

16. . The percent is:
$$23R = 9$$
$$\dfrac{23R}{23} = \dfrac{9}{23}$$
$$R = 0.3913$$
$$R = 39.1\%$$

17. The amount of cashews is:
$$\dfrac{20 \text{ lb}}{6 \text{ lb}} = \dfrac{100 \text{ lb}}{x}$$
$$20x = 600$$
$$x = 30 \text{ lb}$$

18. The percent of no-show ticket holders is:
$$10{,}685R = 475$$
$$\dfrac{10{,}685R}{10{,}685} = \dfrac{475}{10{,}685}$$
$$R = 0.04$$
$$R = 4\%$$

19. The percent of discount is:
$$\$225R = \$67.50$$
$$\dfrac{\$225R}{\$225} = \dfrac{\$67.50}{\$225}$$
$$R = 0.3$$
$$R = 30\%$$

20. The selling price is:
$$P = 175\%(\$255)$$
$$P = 1.75(\$255)$$
$$P = \$446.25$$

CHAPTER EIGHT
EQUATIONS IN TWO VARIABLES

Section 8.1
The Rectangular Coordinate System

1. $(0,-3)$ is a solution
$$2x - y = 3$$
$$2(0) - (-3) = 3$$
$$0 + 3 = 3$$
$$3 = 3$$

3. $(4,-5)$ is not a solution
$$x + y = 9$$
$$4 + (-5) = 9$$
$$-1 \neq 9$$

5. $(3,-4)$ is a solution
$$x + y = -1$$
$$3 + (-4) = -1$$
$$-1 = -1$$

7. $(1,-1)$ is a solution
$$2x - y = 3$$
$$2(1) - (-1) = 3$$
$$2 + 1 = 3$$
$$3 = 3$$

9. $(0,-5)$ is a solution
$$3x - 2y = 10$$
$$3(0) - 2(-5) = 10$$
$$0 + 10 = 10$$
$$10 = 10$$

11. $(-10,7)$ is a solution
$$2x + 4y = 8$$
$$2(-10) + 4(7) = 8$$
$$-20 + 28 = 8$$
$$8 = 8$$

13. $(-4,-12)$ is not a solution
$$3x - 7y = -72$$
$$3(-4) - 7(-12) = -72$$
$$-12 + 84 = -72$$
$$72 \neq -72$$

15. $(-4,9)$ is a solution
$$2x - 7y + 71 = 0$$
$$2(-4) - 7(9) + 71 = 0$$
$$-8 + (-63) + 71 = 0$$
$$0 = 0$$

17. $\left(-\dfrac{1}{2}, -\dfrac{6}{5}\right)$ is not a solution

$$4x - 5y = -4$$

$$4\left(-\dfrac{1}{2}\right) - 5\left(-\dfrac{6}{5}\right) = -4$$

$$-2 + 6 = -4$$

$$4 \neq -4$$

19. $(2.25, 0.6)$ is a solution

$$60x - 40y = 111$$

$$60(2.25) - 40(0.6) = 111$$

$$135 - 24 = 111$$

$$111 = 111$$

21.-29.

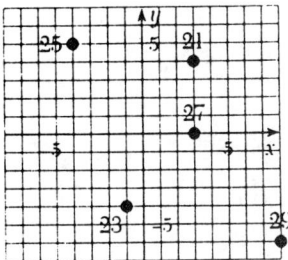

31.- 39.

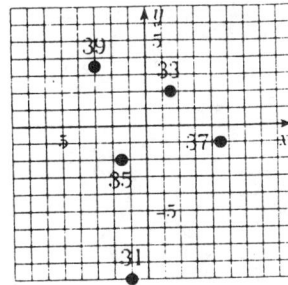

41. A = (3, 4)

43. C = $(2, -4)$

45. E = $(-1, 4)$

47. G = $(-3, 1)$

49. I = $(-7, 6)$

51. A = (4, 1.5)

53. C = $(-3, -0.5)$

55. E = $(4.5, -4)$

57. G = $(2, -6.5)$

59. I = $(0, 1.5)$

61. Yakima = (5, E)

63. Wenatchee = (6, D)

65. Longview

67. Houston = (5, D)

69. Dallas = (5, B)

71. The origin is at x. D is 75 ft below x and is at the same level as bottom of measure to A, so A = 400 ft – 75 ft = 325 ft. The distance from the rock base to E is 325 ft, but the distance from the bridge base to the rock is 75 ft, so 325 ft – 75 ft = 250 ft. D and F are at the same vertical depth. H has the same bottom depth as D and F and The level to the bridge is 75 ft., so H is 250 ft – 75 ft = 175 ft.

73. Begin at the origin and first move 4 units to the left and then move 7 units up.

75. $\left(-\dfrac{27}{5}, -\dfrac{18}{7}\right)$ is not a solution

$$10x - 182y + 167 = 0$$

$$10\left(-\dfrac{27}{5}\right) - 182\left(-\dfrac{18}{7}\right) + 167 = 0$$

$$2(-27) - 26(-18) + 167 = 0$$

$$-54 + 468 + 167 = 0$$

$$581 \neq 0$$

77. answers vary

79. Solve:

$$15y - 4y + 17 = -60$$

$$11y = -77$$

$$y = -7$$

81. Solve:

$$0.35a - 6.7 = 0.25a + 8.93$$

$$0.1a = 15.63$$

$$a = 156.3$$

Section 8.2
Solving Linear Equations in Two Variables

1. The solution is $(6,12)$

$$y = -x + 18$$

$$y = -6 + 18$$

$$y = 12$$

3. The solution is $(13,5)$

$$y = -x + 18$$

$$y = -13 + 18$$

$$y = 5$$

5. The solution is $(-3,21)$

$$y = -x + 18$$

$$y = -(-3) + 18$$

$$y = 21$$

7. The solution is $(8,-9)$

$$y = -\dfrac{1}{4}x - 7$$

$$y = -\dfrac{1}{4}(8) - 7$$

$$y = -2 - 7$$

$$y = -9$$

9. The solution is $(0,-7)$

$$y = -\frac{1}{4}x - 7$$

$$y = -\frac{1}{4}(0) - 7$$

$$y = 0 - 7$$

$$y = -7$$

11. The solution is $(16, -11)$

$$y = -\frac{1}{4}x - 7$$

$$y = -\frac{1}{4}(16) - 7$$

$$y = -4 - 7$$

$$y = -11$$

13. The solution is $(0, 6)$

$$y = \frac{2}{7}x + 6$$

$$y = \frac{2}{7}(0) + 6$$

$$y = 0 + 6$$

$$y = 6$$

15. The solution is $\left(-3, \frac{36}{7}\right)$

$$y = \frac{2}{7}x + 6$$

$$y = \frac{2}{7}(-3) + 6$$

$$y = -\frac{6}{7} + 6 \cdot \frac{7}{7}$$

$$y = -\frac{6}{7} + \frac{42}{7}$$

$$y = \frac{36}{7}$$

17. The solution is $\left(-15, \frac{12}{7}\right)$

$$y = \frac{2}{7}x + 6$$

$$y = \frac{2}{7}(-15) + 6$$

$$y = -\frac{30}{7} + 6 \cdot \frac{7}{7}$$

$$y = -\frac{30}{7} + \frac{42}{7}$$

$$y = \frac{12}{7}$$

19. The solution is $(0,10)$

$$y = 0.8x + 10$$

$$y = 0.8(0) + 10$$

$$y = 0 + 10$$

$$y = 10$$

21. The solution is $(-1.5, 8.8)$

$$y = 0.8x + 10$$

$$y = 0.8(-1.5) + 10$$

$$y = -1.2 + 10$$

$$y = 8.8$$

23. The solution is $(-12.5, 0)$

$$y = 0.8x + 10$$

$$y = 0.8(-12.5) + 10$$

$$y = -10 + 10$$

$$y = 0$$

25. The value after 4 years:

$y = \$45{,}000 - \$3600x$

$y = \$45{,}000 - \$3600(4)$

$y = \$45{,}000 - \$14{,}400$

$y = \$30{,}600$

The value after 8 years:

$y = \$45{,}000 - \$3600x$

$y = \$45{,}000 - \$3600(8)$

$y = \$45{,}000 - \$28{,}800$

$y = \$16{,}200$

27. The second number is:

$x = 2y - 2$

$47 = 2y - 2$

$49 = 2y$

$24.5 = y$

29. The solution is $(3, 12)$

$$y = \frac{3}{2}x + \frac{15}{2}$$

$$y = \frac{3}{2}(3) + \frac{15}{2}$$

$$y = \frac{9}{2} + \frac{15}{2}$$

$$y = \frac{24}{2}$$

$$y = 12$$

31. The solution is $\left(-\dfrac{16}{3}, -\dfrac{1}{2}\right)$

$$y = \frac{3}{2}x + \frac{15}{2}$$

$$y = \frac{3}{2}\left(-\frac{16}{3}\right) + \frac{15}{2}$$

$$y = \frac{-16}{2} + \frac{15}{2}$$

$$y = -\frac{1}{2}$$

33. The solution is $(-4.2, 1.2)$

$$y = \frac{3}{2}x + \frac{15}{2}$$

$$y = \frac{3}{2}(-4.2) + \frac{15}{2}$$

$$y = 1.5(-4.2) + 7.5$$

$$y = -6.3 + 7.5$$

$$y = 1.2$$

35. The number of 6 lb. boxes is:

$129 = 6x + 7y$

$129 = 6x + 7(9)$

$129 = 6x + 63$

$66 = 6x$

$11 = x$

37. The solution is $(3, 0)$.

$$y = -\frac{7}{3}x + 7$$

$$y = -\frac{7}{3}(3) + 7$$

$$y = -7 + 7$$

$$y = 0$$

39. The solution is $\left(-5, \frac{56}{3}\right)$.

$$y = -\frac{7}{3}x + 7$$

$$y = -\frac{7}{3}(-5) + 7 \cdot \frac{3}{3}$$

$$y = \frac{35}{3} + \frac{21}{3}$$

$$y = \frac{56}{3}$$

41. The solution is $\left(\frac{2}{7}, \frac{19}{3}\right)$.

$$y = -\frac{7}{3}x + 7$$

$$y = -\frac{7}{3}\left(\frac{2}{7}\right) + 7 \cdot \frac{3}{3}$$

$$y = -\frac{2}{3} + \frac{21}{3}$$

$$y = \frac{19}{3}$$

43. The no. of servings of Cheerios is:

$$2(28) + 23y = 171$$

$$56 + 23y = 171$$

$$23y = 115$$

$$y = 5$$

45. The length of side "a" is:
$$P = 2a + b$$

$$38\,cm = 2a + 10\,cm$$

$$28\,cm = 2a$$

$$14\,cm = a$$

47. 8 subscriptions provides:

$$T = 21.75s$$

$$T = 21.75(8)$$

$$T = \$174$$

25 subscriptions provides:

$$T = 21.75s$$

$$T = 21.75(25)$$

$$T = \$543.75$$

1 subscriptions provides:

$$T = 21.75s$$

$$T = 21.75(1)$$

$$T = \$21.75$$

49. The solution is $(1, 500)$.
$$I = 500t$$
$$I = 500(1)$$
$$I = 500$$

The solution is $(2, 1000)$.
$$I = 500t$$
$$I = 500(2)$$
$$I = 1000$$

The solution is $(5, 2500)$.
$$I = 500t$$
$$I = 500(5)$$
$$I = 2500$$

The solution is $(10, 5000)$.
$$I = 500t$$
$$I = 500(10)$$
$$I = 5000$$

51. The solution is $(1, 1000)$
$$V = 1000t$$
$$V = 1000(1)$$
$$V = 1000$$

The solution is $(2, 2000)$
$$V = 1000t$$
$$V = 1000(2)$$
$$V = 2000$$

The solution is $(3, 3000)$
$$V = 1000t$$
$$V = 1000(3)$$
$$V = 3000$$

The solution is $(6, 6000)$
$$V = 1000t$$
$$V = 1000(6)$$
$$V = 6000$$

The solution is $(10, 10,000)$
$$V = 1000t$$
$$V = 1000(10)$$
$$V = 10,000$$

53. The solution for the problem is $(4, 5)$.
The x value is stated first and the y value second.

55. The solution is $\left(4, -\dfrac{2}{5}\right)$
$$5y + 2 = 0$$
$$5y = -2$$
$$y = -\dfrac{2}{5}$$

57. The equation with solutions of $(3, 0)$ and $(0, 3)$ is $x + y = 3$ or $y = -x + 3$

59. The equation with solutions of $(6, 0)$ and $(0, -3)$ is $x - 2y = 6$ or $y = \dfrac{x}{2} - 3$

61. $y = 0.05$ is not a solution
$$2y - 13.5 \neq 12.6 - 16y$$
$$2(0.05) - 13.5 \approx 12.6 - 16(0.05)$$
$$0.1 - 13.5 \neq 12.6 - 0.8$$
$$-13.4 \neq 11.8$$

63. $t = -4c - 3d + 5 = -4(6) - 3(-2) + 5 = -24 + 6 + 5 = -13$

Section 8.3
Graphing Linear Equations in Two Variables

1. $y = \dfrac{1}{2}x$

x	y
−6	−3
0	0
6	3

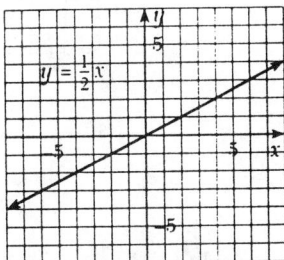

3. $y = -x - 2$

x	y
0	−2
−2	0
1	−3

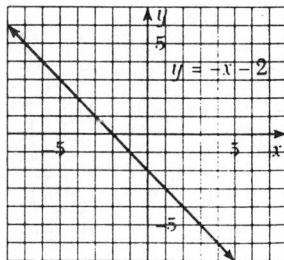

5. $y = x - 4$

x	y
0	−4
4	0
2	−2

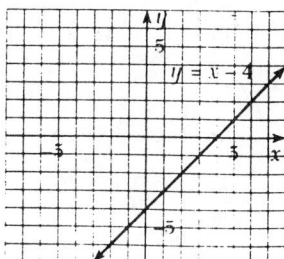

7. $y = -x - 9$

x	y
-3	-6
-4	-5
-6	-3

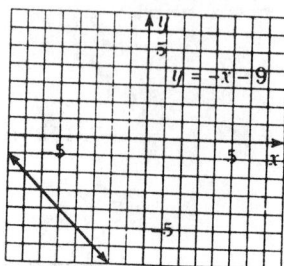

9. $y = 2x + 1$

x	y
4	9
-2	-3
2	5

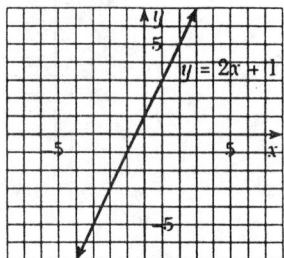

11. The graph of $y = -\dfrac{1}{2}x + 3$ is:

13. The graph of $y = 2x + 4$ is:

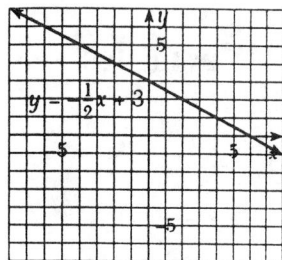

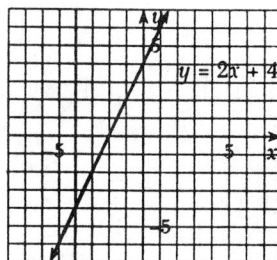

15. The graph of $y = -\dfrac{3}{4}x + 3$ is:

17. The graph of $y = \dfrac{4}{5}x - 4$ is:

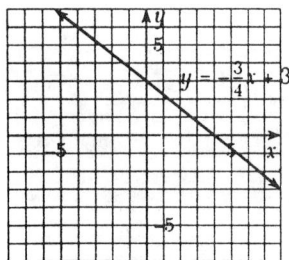

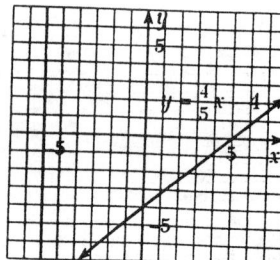

19. The graph of $y = -\dfrac{3}{5}x - 3$ is:

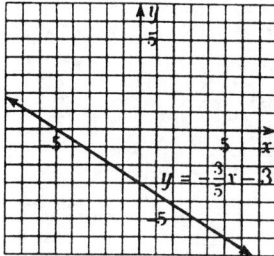

21. The graph of $y = \dfrac{4}{3}x - 4$ is:

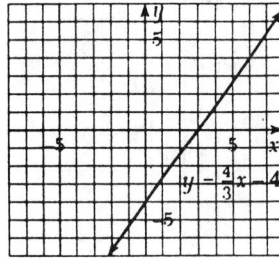

23. The graph of $y = \dfrac{1}{2}x + 3$ is:

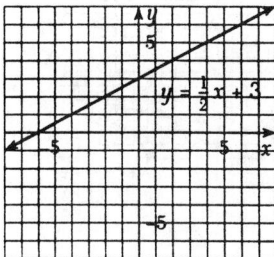

25. The graph of $y = \dfrac{5}{7}x - \dfrac{10}{7}$ is:

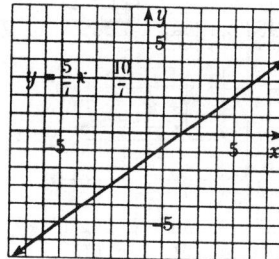

27. The graph of $y = 0.6x - 1.2$ is:

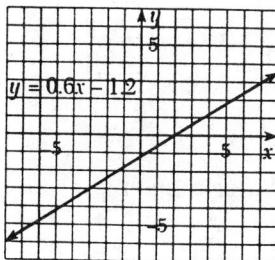

29. The graph of $y = -\dfrac{7}{4}x + 7$ is:

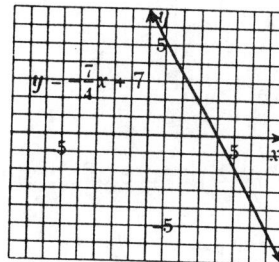

31. The graph of $y = -1500x + 8000$ is:

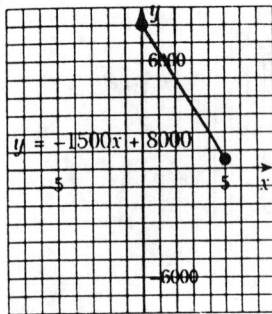

33. The graph of $0.25x + 0.30v = 300$ is:

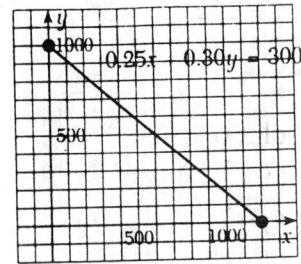

35. The graph of $S = 740M$ is:

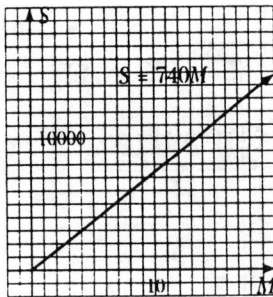

37. The graph of $P = 40 + 2w$ is:

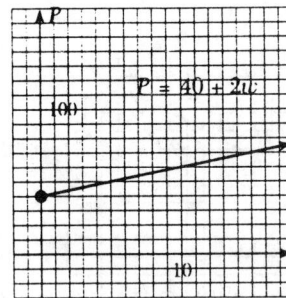

39. answers vary

41. answers vary

43. A line graph is a picture of all of the ordered pairs that render the equation true.

45. The graph of $2y - 12 = 0$ is:

47. group activity

49. The triangle with sides of 28 cm, 42 cm and 16 cm is not a right triangle.

$$a^2 + b^2 = c^2$$
$$28^2 + 16^2 = 42^2$$
$$784 + 256 = 1764$$
$$1040 \neq 1764$$

51. The solution is $(-4, 2)$

$$4x - 7y = -30$$
$$4(-4) - 7y = -30$$
$$-16 - 7y = -30$$
$$-7y = -14$$
$$y = 2$$

Section 8.4
Slopes and Intercepts (Optional)

1. The x-intercept is $(4, 0)$ and the y-intercept is $(0, -6)$.

3. The x-intercept is $(-7, 0)$ and the y-intercept is $(0, 1)$.

5. The x-intercept is $(4, 0)$.

$$4x + y = 16$$
$$4x + 0 = 16$$
$$4x = 16$$
$$x = 4$$

The y-intercept is $(0, 16)$.

$$4x + y = 16$$
$$4(0) + y = 16$$
$$0 + y = 16$$
$$y = 16$$

7. The x-intercept is $(9, 0)$.

$$2x + 9y = 18$$
$$2x + 9(0) = 18$$
$$2x + 0 = 18$$
$$2x = 18$$
$$x = 9$$

The y-intercept is $(0, 2)$.

$$2x + 9y = 18$$
$$2(0) + 9y = 18$$
$$0 + 9y = 18$$
$$9y = 18$$
$$y = 2$$

9. The *x*-intercept is (6, 0).

$$8x - 3y = 48$$
$$8x - 0 = 48$$
$$8x = 48$$
$$x = 6$$

The *y*-intercept is (0, −16).

$$8x - 3y = 48$$
$$8(0) - 3y = 48$$
$$0 - 3y = 48$$
$$-3y = 48$$
$$y = -16$$

11. The *x*-intercept is (7.5, 0) and the *y*-intercept is (0, −2.5).

13. The *x*-intercept is (1.5, 0) and the *y*-intercept is (0, 6.5).

15. The *x*-intercept is (5, 0).

$$-3x + 2y = -15$$
$$-3x + 2(0) = -15$$
$$-3x + 0 = -15$$
$$-3x = -15$$
$$x = 5$$

The *y*-intercept is $(0, -\dfrac{15}{2})$.

$$-3x + 2y = -15$$
$$-3(0) + 2y = -15$$
$$0 + 2y = -15$$
$$2y = -15$$
$$x = -\dfrac{15}{2}$$

17. The *x*-intercept is (21, 0).

$$y = -\frac{1}{3}x + 7$$
$$0 = -\frac{1}{3}x + 7$$
$$-7 = -\frac{1}{3}x$$
$$21 = x$$

The *y*-intercept is (0, 7).

$$y = -\frac{1}{3}x + 7$$
$$y = -\frac{1}{3}(0) + 7$$
$$y = 0 + 7$$
$$y = 7$$

19. The *x*-intercept is $(-\dfrac{8}{7}, 0)$.

$$4y = -7x - 8$$
$$4(0) = -7x - 8$$
$$0 = -7x - 8$$
$$8 = -7x$$
$$-\frac{8}{7} = x$$

The *y*-intercept is (0, −2).

$$4y = -7x - 8$$
$$4y = -7(0) - 8$$
$$4y = 0 - 8$$
$$4y = -8$$
$$y = -2$$

21. Two points on the line are (−6, 0) and (5, −3).

$$\text{The slope is } \frac{y_2 - y_1}{x_2 - x_1} = \frac{-3 - 0}{5 - (-6)} = \frac{-3}{5 + 6} = \frac{-3}{11}$$

23. Two points on the line are (−7, 4) and (0, 3).

$$\text{The slope is } \frac{y_2 - y_1}{x_2 - x_1} = \frac{3 - 4}{0 - (-7)} = \frac{-1}{0 + 7} = \frac{-1}{7}$$

25. $$\text{The slope is } \frac{y_2 - y_1}{x_2 - x_1} = \frac{7 - 0}{0 - 3} = -\frac{7}{3}$$

27. $$\text{The slope is } \frac{y_2 - y_1}{x_2 - x_1} = \frac{7 - 6}{3 - 4} = \frac{1}{-1} = -1$$

29. $$\text{The slope is } \frac{y_2 - y_1}{x_2 - x_1} = \frac{2 - (-1)}{-10 - 6} = \frac{2 + 1}{-16} = \frac{3}{-16}$$

31. Two points on the line are (0, 5) and (−6, 3).

$$\text{The slope is } \frac{y_2 - y_1}{x_2 - x_1} = \frac{3 - 5}{-6 - 0} = \frac{-2}{-6} = \frac{1}{3}$$

33. Two points on the line are (1, −6) and (−8, 6).

$$\text{The slope is } \frac{y_2 - y_1}{x_2 - x_1} = \frac{6 - (-6)}{-8 - (-3)} = \frac{6 + 6}{-8 + 3} = \frac{12}{-5}$$

35. $$\text{The slope is } \frac{y_2 - y_1}{x_2 - x_1} = \frac{-4 - (-1)}{5 - (-2)} = \frac{-4 + 1}{5 + 2} = \frac{-3}{7}$$

37. $$\text{The slope is } \frac{y_2 - y_1}{x_2 - x_1} = \frac{5 - 1}{-2 - 5} = \frac{4}{-7}$$

39. $$\text{The slope is } \frac{y_2 - y_1}{x_2 - x_1} = \frac{-2 - 3}{-12 - 0} = \frac{-5}{-12} = \frac{5}{12}$$

41. The y-intercept is $(0, 2)$.   The x-intercept is $\left(\dfrac{14}{3}, 0\right)$

$$3x + 7y = 14$$
$$3(0) + 7y = 14$$
$$0 + 7y = 14$$
$$y = 2$$

$$3x + 7y = 14$$
$$3x + 7(0) = 14$$
$$3x + 0 = 14$$
$$x = \dfrac{14}{3}$$

The slope is $\dfrac{y_2 - y_1}{x_2 - x_1} = \dfrac{2 - 0}{0 - \dfrac{14}{3}} = \dfrac{2}{-\dfrac{14}{3}} = \dfrac{2}{-\dfrac{14}{3}} \cdot \dfrac{3}{3} = \dfrac{6}{-14} = -\dfrac{3}{7}$

43. The y-intercept is $\left(0, -\dfrac{9}{4}\right)$.   The x-intercept is $(-3, 0)$.

$$-3x - 4y = 9$$
$$-3(0) - 4y = 9$$
$$0 - 4y = 9$$
$$y = -\dfrac{9}{4}$$

$$-3x - 4y = 9$$
$$-3x - 4(0) = 9$$
$$-3x + 0 = 9$$
$$x = -3$$

The slope is $\dfrac{y_2 - y_1}{x_2 - x_1} = \dfrac{-\dfrac{9}{4} - 0}{0 - (-3)} = \dfrac{-\dfrac{9}{4}}{3} = \dfrac{-\dfrac{9}{4}}{3} \cdot \dfrac{4}{4} = \dfrac{-9}{12} = -\dfrac{3}{4}$

45. The y-intercept is $\left(0, -\dfrac{7}{2}\right)$.   The x-intercept is $\left(\dfrac{7}{8}, 0\right)$

$$8x - 2y = 7$$
$$8(0) - 2y = 7$$
$$0 - 2y = 7$$
$$y = -\dfrac{7}{2}$$

$$8x - 2y = 7$$
$$8x - 2(0) = 7$$
$$8x + 0 = 7$$
$$x = \dfrac{7}{8}$$

The slope is $\dfrac{y_2 - y_1}{x_2 - x_1} = \dfrac{-\dfrac{7}{2} - 0}{0 - \dfrac{7}{8}} = \dfrac{-\dfrac{7}{2}}{-\dfrac{7}{8}} = \dfrac{-\dfrac{7}{2}}{-\dfrac{7}{8}} \cdot \dfrac{8}{8} = \dfrac{-28}{-7} = 4$

47. The pitch of the roof is $\dfrac{y_2 - y_1}{x_2 - x_1} = \dfrac{1.5\,\text{ft}}{2.5\,\text{ft}} \cdot \dfrac{10}{10} = \dfrac{15}{25} = \dfrac{3}{5}$

49. The slope of the ramp is $\dfrac{y_2 - y_1}{x_2 - x_1} = \dfrac{38\,\text{ft}}{27\,\text{ft}} = 1.407$

51. The slope of the ramp is $\dfrac{y_2 - y_1}{x_2 - x_1} = \dfrac{84\,\text{ft}}{66\,\text{ft}} = \dfrac{14}{11} = 1.27$

53. The grandparents invest $500 each year into the college fund.
    The slope of the line between (1, 500) and (5, 2500) is:
    $$\dfrac{y_2 - y_1}{x_2 - x_1} = \dfrac{2500 - 500}{5 - 1} = \dfrac{2000}{4} = 500$$

55. The slope of the line is negative.

57. The slope of a line can be determined by reading off the x and y intercepts from the graph of the line.

59. The slope is $\dfrac{y_2 - y_1}{x_2 - x_1} = \dfrac{-5 - 7}{7 - 7} = \dfrac{-12}{0} =$ undefined

61. group activity　　　　　　63. $3.45 = 345\%$　　　　　　65. $113.8\% = 1.138$

Section 8.5
Variation (Optional)

1. $d = kr$ or $d = rt$

3. $P = ks$

5. $C = kn$

7. m goes from 20 to 60

9. The constant of variation is:
    $$y = kx$$
    $$24 = 16k$$
    $$k = \dfrac{3}{2} = 1.5$$

11. The constant of variation is:
    $$y = kx^2$$
    $$14 = 7^2 k$$
    $$k = \dfrac{14}{49} = \dfrac{2}{7}$$

13. The constant of variation is:

$$a = kb$$
$$14 = 6k$$
$$k = \frac{14}{6} = \frac{7}{3}$$

When $b = 16$, then a is:

$$a = kb$$
$$a = \frac{7}{3} \cdot 16$$
$$a = \frac{112}{3} = 37\frac{1}{3}$$

15. The constant of variation is:

$$m = kn^2$$
$$30 = 5^2 k$$
$$30 = 25k$$
$$k = \frac{30}{25} = \frac{6}{5} = 1.2$$

When $n = 16$, then m is:

$$m = kn^2$$
$$m = 16^2 \cdot \frac{6}{5}$$
$$m = 256 \cdot \frac{6}{5}$$
$$m = \frac{1536}{5} = 307\frac{1}{5}$$

17. $lw = k$

19. $Cp = k$

21. $ns = k$

23. B goes from 20 to 40

25. The constant of variation is:

$$yx = k$$
$$108 \cdot \frac{2}{3} = k$$
$$k = 72$$

27. The constant of variation is:

$$yx = k$$
$$24(0.5) = k$$
$$k = 12$$

29. The constant of variation is:

$$yx = k$$
$$42(16) = k$$
$$k = 672$$

When $x = 48$, then y is:

$$yx = k$$
$$y(48) = 672$$
$$y = \frac{672}{48}$$
$$y = 14$$

31. The constant of variation is:

$$yx = k$$
$$22(4) = k$$
$$k = 88$$

When $x = 8$, then y is:

$$yx = k$$
$$y(8) = 88$$
$$y = \frac{88}{8}$$
$$y = 11$$

33. The constant of variation is:

$$w = vk$$

$$350 = \frac{5}{3}k$$

$$350 \cdot \frac{3}{5} = \frac{5}{3} \cdot \frac{3}{5}k$$

$$k = 70 \cdot 3$$

$$k = 210$$

When $v = \frac{7}{3}$, then w is:

$$w = vk$$

$$w = \frac{7}{3}(210)$$

$$w = 7(70)$$

$$w = 490 \text{ lb}$$

35. The constant of variation is:

$$s = tk$$

$$372 = 3100k$$

$$\frac{372}{3100} = k$$

$$k = 0.12$$

When $t = \$5400$, then s is:

$$s = tk$$

$$s = \$5400(0.12)$$

$$s = \$648$$

37. The constant of variation is:

$$a = wk$$

$$\frac{5}{11} = 50k$$

$$\frac{5}{11} \cdot \frac{1}{50} = 50 \cdot \frac{1}{50}k$$

$$\frac{1}{110} = k$$

When $w = 75$, then a is:

$$a = wk$$

$$a = 75 \cdot \frac{1}{110}$$

$$a = \frac{75}{110} = \frac{15}{22} \text{ amps}$$

39. The constant of variation is:

$$lw = A$$

$$(12)8 = A$$

$$96 = A$$

When $w = 6$, then l is:

$$lw = A$$

$$6l = 96$$

$$l = \frac{96}{6} = 16 \text{ ft}$$

41. The constant of variation is:

$$p = ks$$

$$46,250 = 18,500k$$

$$\frac{46,250}{18,500} = k$$

$$2.5 = k$$

When $s = 24,000$, then p is:

$$p = ks$$

$$p = 2.5(24,000)$$

$$p = \$60,000$$

43. The amount of money invested varies directly as the time of the investment. The longer the time, the more the investment: $A = 500t$

45. answers vary

47. The constant of variation is:

When $r = 100$ km/hr, then t is:

$$tr = d$$

$$tr = d$$
$$(2.5)55 = d$$
$$137.5 = d$$

$$\frac{100 \text{ km}}{\text{hr}} \cdot \frac{\text{mi}}{1.609 \text{ km}} t = 137.5$$
$$62.15t = 137.5$$
$$t = \frac{137.5}{62.15}$$
$$t = 2.2 \text{ hr}$$

49. $\dfrac{-21.06t^2}{-1.56t} = 13.5t$

51. $\dfrac{9r^2}{0.5s} = \dfrac{9(2.3)^2}{0.5(-1.8)} = \dfrac{9(5.29)}{(-0.9)} = -52.9$

53. -3.5 is not a solution
$$0.8 = 31.2z + 100$$
$$0.8 = 31.2(-3.5) + 100$$
$$0.8 = -109.2 + 100$$
$$0.8 \neq -9.2$$

Chapter 8
True-False Concept Review

1. False – Equations with two variables sometimes have no solution.

2. False – The ordered pair (2, 3) is not a solution of the equation $2y - x = 1$

3. True

4. True

5. True

6. False – The quadrants of the rectangular coordinate system are numbered counterclockwise.

7. False – The point where the x-axis and the y-axis intersect is called the "origin."

8. False – The x-value of an ordered pair shows the distance of a point from the y-axis.

9. True

10. False – If either x or y is zero in an ordered pair, then the point corresponding to the pair lies on one of the axes.

11. False – For all values of a and b that are not both equal to zero, the graph of $ax + by = c$ is a straight line.

12. False – At least two points are needed to determine the position of a line on a graph.

13. True

14. False – Line graphs are extremely useful in practical applications.

15. False – The point $(-2,-1)$ is in quadrant III.

16. True

17. True

18. False – The slope of a line that goes up from right to left is negative.

19. True

20. True

Chapter 8
Review

1. $(4,-1)$ is a solution
$$3x - 8y = 20$$
$$3(4) - 8(-1) = 20$$
$$12 + 8 = 20$$
$$20 = 20$$

3. $(-1,-1)$ is a solution
$$4x - 7y = 3$$
$$4(-1) - 7(-1) = 3$$
$$-4 + 7 = 3$$
$$3 = 3$$

5. $(-3,-8)$ is not a solution
$$7x + 5y = 19$$
$$7(-3) + 5(-8) = 19$$
$$-21 - 40 = 19$$
$$-61 \neq 19$$

7 – 9. The points $(4, -1)$ and $\left(\frac{1}{2},-6\right)$ are plotted as follows:

7.
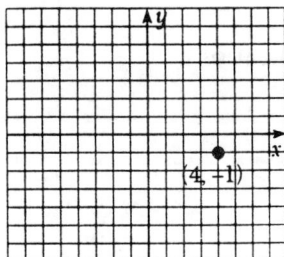

9.
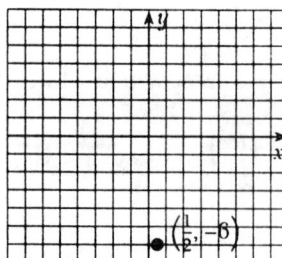

11. $A = (6,-3)$

13. $C = (4, 5.5)$

15. $E = (-1.5, -2)$
$$y = \frac{4}{5}x - 4$$
$$y = \frac{4}{5}(5) - 4$$
$$y = 4 - 4$$
$$y = 0$$

17. The solution is $(10, 4)$
$$y = \frac{4}{5}x - 4$$
$$y = \frac{4}{5}(10) - 4$$
$$y = 4(2) - 4$$
$$y = 8 - 4$$
$$y = 4$$

19. The solution is $(14, 6)$ 21.

$$y = \frac{6}{7}x - 6$$

$$6 = \frac{6}{7}x - 6$$

$$12 = \frac{6}{7}x$$

$$12 \cdot \frac{7}{6} = \frac{6}{7} \cdot \frac{7}{6}x$$

$$14 = x$$

21. The graph of $y = 3x - 6$ is:

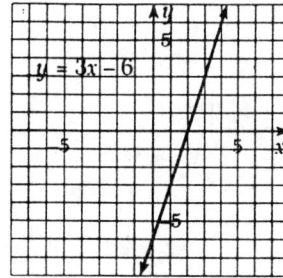

23. The graph of $y = -\frac{2}{3}x - 2$

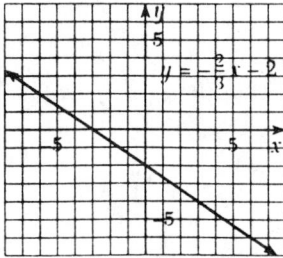

25. The graph of $y = \frac{1}{6}x - \frac{1}{2}$ is:

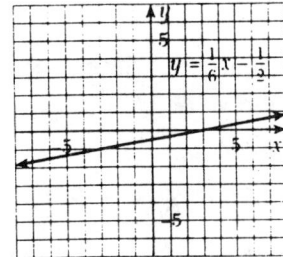

27. The x-intercept is $(8, 0)$

$$3x - 8y = 24$$

$$3x - 8(0) = 24$$

$$3x - 0 = 24$$

$$3x = 24$$

$$x = 8$$

The y-intercept is $(0, -3)$.

$$3x - 8y = 24$$

$$3(0) - 8y = 24$$

$$0 - 8y = 24$$

$$-8y = 24$$

$$y = -3$$

29. The x-intercept is $\left(\frac{5}{2}, 0\right)$

$$6x + 7y = 15$$

$$6x + 7(0) = 15$$

$$6x + 0 = 15$$

$$6x = 15$$

$$x = \frac{15}{6} = \frac{5}{2}$$

The y-intercept is $\left(0, \frac{15}{7}\right)$.

$$6x + 7y = 15$$

$$6(0) + 7y = 15$$

$$0 + 7y = 15$$

$$7y = 15$$

$$y = \frac{15}{7}$$

31. The two points are (3, 1) and (–5, 4).

The slope is $\dfrac{y_2 - y_1}{x_2 - x_1} = \dfrac{4-1}{-5-3} = \dfrac{3}{-8}$

33. The slope is $\dfrac{y_2 - y_1}{x_2 - x_1} = \dfrac{3-(-2)}{4-5} = \dfrac{3+2}{-1} = \dfrac{5}{-1} = -5$

35. The slope is $\dfrac{y_2 - y_1}{x_2 - x_1} = \dfrac{-13-14}{9-21} = \dfrac{-27}{-12} = \dfrac{9}{4}$

37. The constant of variation is:

$s = kt$

$24 = 6k$

$4 = k$

When $t = 96$ mi, then s is:

$s = kt$

$s = 4(96)$

$s = 384$

39. The constant of variation is:

$w = vk$

$1225.5 = 4.3k$

$285 = k$

When $v = 2.8$, then w is:

$w = vk$

$w = 285(2.8)$

$w = 798\,\text{lb}$

41. The constant of variation is:

$yx = k$

$40(8) = k$

$320 = k$

43. The constant of variation is:

$ab = k$

$45(15) = k$

$675 = k$

When $b = 75$, then a is:

$ab = k$

$75a = 675$

$a = 9$

45. The constant of variation is:

$dn = k$

$25(250) = k$

$6250 = k$

When $n = 200$, then d is:

$dn = k$

$200d = 6250$

$d = \$31.25$

Chapter 8
Test

1. If $x = -25$, then y is:

$$y = \frac{3}{5}x - 9$$

$$y = \frac{3}{5}(-25) - 9$$

$$y = -15 - 9$$

$$y = -24$$

2. $(-7, 4)$ is a solution

$$3x + 7y = 7$$

$$3(-7) + 7(4) = 7$$

$$-21 + 28 = 7$$

$$7 = 7$$

3. The graph of the points is:

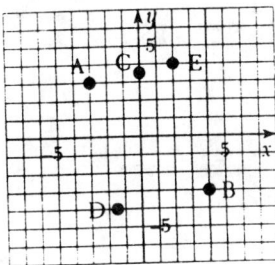

4. The graph of $y = 2x - 6$ is:

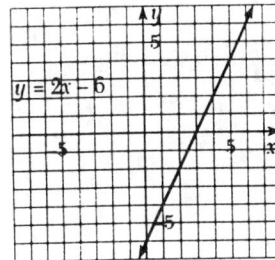

5. $(-4, -8)$ is not a solution.

$$4y - 7x = 40$$

$$4(-8) - 7(-4) = 40$$

$$-32 + 28 = 40$$

$$-4 \neq 40$$

6. $A = (7, 4)$

$B = (-5, 1)$

$C = (3, -2)$

$D = (-3.5, -4)$

$E = (2.5, 5)$

7. The graph of $y = -\frac{1}{2}x + 3$

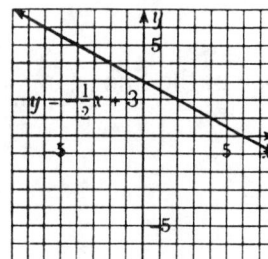

8. If $x = -2$, then y is:

$$y = \frac{1}{7}x - \frac{12}{7}$$

$$y = \frac{1}{7}(-2) - \frac{12}{7}$$

$$y = -\frac{2}{7} - \frac{12}{7}$$

$$y = -\frac{14}{7} = -2$$

9. The graph of the points is:

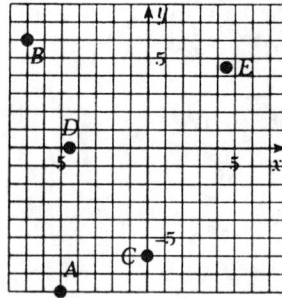

10. If $x = 44$, then y is:

$$y = -\frac{1}{11}x + 11$$

$$y = -\frac{1}{11}(44) + 11$$

$$y = -4 + 11$$

$$y = 7$$

11. $(-5.5, 0.4)$ is a solution

$$y = -3x - 16.1$$

$$0.4 = -3(-5.5) - 16.1$$

$$0.4 = 16.5 - 16.1$$

$$0.4 = 0.4$$

12. The graph of $y = -4x + 6$ is:

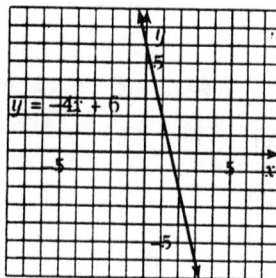

13. The x-intercept is $(5, 0)$

The y-intercept is $(0, -5.5)$

14. The slope of the line is: $\dfrac{y_2 - y_1}{x_2 - x_1} = \dfrac{4 - 0}{6 - 0} = \dfrac{4}{6} = \dfrac{2}{3}$

15. The *x*-intercept is (6, 0).

$$3x - 7y = 18$$
$$3x - 7(0) = 18$$
$$3x - 0 = 18$$
$$3x = 18$$
$$x = 6$$

The *y*-intercept is $(0, -\frac{18}{7})$.

$$3x - 7y = 18$$
$$3(0) - 7y = 18$$
$$0 - 7y = 18$$
$$-7y = 18$$
$$x = -\frac{18}{7}$$

16. The slope of the line is: $\dfrac{y_2 - y_1}{x_2 - x_1} = \dfrac{5 - (-7)}{-3 - 6} = \dfrac{5 + 7}{-9} = \dfrac{12}{-9} = -\dfrac{4}{3}$

17. The constant of variation is:

$$y = kx$$
$$9 = 40k$$
$$\frac{9}{40} = k$$

When $x = 90$, then y is:

$$y = kx$$
$$y = \frac{9}{40}(90)$$
$$y = \frac{81}{4}$$

18. The constant of variation is:

$$yx = k$$
$$18(27) = k$$
$$486 = k$$

When $x = 45$, then y is:

$$yx = k$$
$$45y = 486$$
$$y = 10.8$$

19. The graph of $C = 20 + 0.25m$ is:

The cost is:

$$C = 20 + 0.25m$$
$$C = 20 + 0.25(325)$$
$$C = 20 + 81.25$$
$$C = \$101.25$$

CHAPTERS 1 - 8
CUMULATIVE REVIEW

1. The average is: $\dfrac{3456 + 7234 + 8110 + 1375 + 9120}{5} = \dfrac{29295}{5} = 5859$

3. $(9 \cdot 2)^5 = 9^5 \cdot 2^5$

5. $297,987 \times 10^6 = 297,987,000,000$

7. $C = \pi d = 3.14(8) = 25.12$

9. 12 is a solution
$$8x - 15 = 81$$
$$8(12) - 15 = 81$$
$$96 - 15 = 81$$
$$81 = 81$$

11. $\$215,000 - \$253,000 = -\$38,000$

13. Simplify:
$$\frac{4^2 - 4[6 + 3(-4)]^2}{(-14 + 16)^2} = \frac{16 - 4[6 + (-12)]^2}{(2)^2} = \frac{16 - 4[-6]^2}{4} =$$
$$\frac{16 - 4(36)}{4} = \frac{16 - 144}{4} = \frac{-128}{4} = -32$$

15. The number is:
$$6 - 8 + 4n = 10 + 2n$$
$$-2 + 4n = 10 + 2n$$
$$2n = 12$$
$$n = 6$$

17. Add:

$$-\frac{2x}{9}+\frac{7x}{12}+\frac{x}{4}$$

$$-\frac{2x}{9}\cdot\frac{4}{4}+\frac{7x}{12}\cdot\frac{3}{3}+\frac{x}{4}\cdot\frac{9}{9}$$

$$-\frac{8x}{36}+\frac{21x}{36}+\frac{9x}{36}\quad:$$

$$\frac{22x}{36}$$

$$\frac{11x}{18}$$

19. $-\dfrac{4}{3}$ is a solution

$$\frac{5}{2}-\frac{3x}{8}=3$$

$$\frac{5}{2}-\frac{3\left(-\dfrac{4}{3}\right)}{8}=3$$

$$\frac{5}{2}-\frac{-4}{8}=3$$

$$\frac{5}{2}\cdot\frac{4}{4}+\frac{4}{8}=3$$

$$\frac{20}{8}+\frac{4}{8}=3$$

$$\frac{24}{8}=3$$

$$3=3$$

21. Solve:
$$0.8x-0.024=16$$
$$0.8x=16.024$$
$$x=20.03$$

23. The average is: $\dfrac{22+20+18}{3}=\dfrac{60}{3}=20$

25. $\dfrac{1}{4}\cdot\dfrac{37}{2}=4\dfrac{5}{8}\,\text{gal}$

27. The watch will be:
$$\frac{0.4}{5} = \frac{x}{72}$$
$$5x = 0.4(72)$$
$$5x = 28.8$$
$$x = 5.76 \text{ min fast}$$

29. The number of correct answers is:
$$70\%(148) = x$$
$$0.7(148) = x$$
$$103.6 = x$$
$$104 \approx x$$

31. The ordered pair is $(-2,-14)$
$$7x - 2y = 14$$
$$7(-2) - 2y = 14$$
$$-14 - 2y = 14$$
$$-2y = 28$$
$$y = -14$$

33. The ordered pair is $(2, 0)$
$$7x - 2y = 14$$
$$7x - 2(0) = 14$$
$$7x - 0 = 14$$
$$7x = 14$$
$$x = 2$$

35. The graph of $3x - 2y = 2$ is:

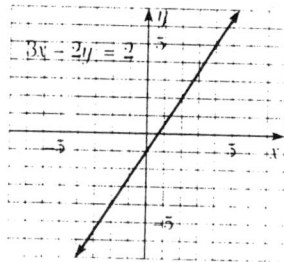

37. The difference between the no. of compact rentals and the total of all others was:
$$3972 - 2338 = 1634 \qquad 2338 - 1634 = 704$$

39. The percent of the rentals that are compacts and standards is:
$$\frac{2338 + 1147}{3972} = \frac{3485}{3972} = 0.88 = 88\%$$

41. The occupancy rate was the highest in 1992.

43. The occupancy rate increase between 1990 and 1991 was 13%. Therefore, the occupancy rate will be 93% in 1994.

45. The number of vacant apartments in 1992 was $0.05(1500) = 75$

47. Subtract:
3 hr 25 min 41 sec = 2 hr 84 min 101 sec
−1 hr 30 min 54 sec = 1 hr 30 min 54 sec
⎯⎯⎯⎯⎯⎯⎯⎯⎯⎯⎯⎯⎯⎯⎯⎯⎯⎯
1 hr 54 min 47 sec

234

49. The average is $\dfrac{(12+22+35+28+45+60)\text{lb}}{6} = \dfrac{(202)\text{lb}}{6} = 33\dfrac{2}{3}\text{lb}$

51. The speed in ft/sec is $\dfrac{320\text{ mi}}{\text{hr}} \cdot \dfrac{5280\text{ ft}}{\text{mi}} \cdot \dfrac{\text{hr}}{3600\text{ sec}} = \dfrac{469\dfrac{1}{3}\text{ ft}}{\text{sec}}$

53. The perimeter is $[2(18)+2(12)+2(4)+2(9)]\text{ft} = [36+24+8+18]\text{ft} = 86\text{ ft}$

   The area of the deck is $(16\text{ ft})(21\text{ ft}) = 336$ sq. ft.

   The gallons of stain needed is $336\text{ ft}^2 \div 125\dfrac{\text{gal}}{\text{ft}^2} = 2.68\text{ gal} \Rightarrow 3\text{ gal}$

55. The volume is $\pi r^2 h = 3.14\left(\dfrac{21}{2}\text{ in}\right)^2 (30\text{ in}) = 10385.55\text{ in}^3 \approx 10386\text{ in}^3$

57. The coordinates of the points are:
   $A = (-4,7)$
   $B = (3,6)$
   $C = (-2,5)$
   $D = (0,-1)$
   $E = (6,-4)$

Section G.1
Reasoning

1. 1, 4, 9, 16, 25, <u>36</u>

3. 5, 9, 13, 17, 21, <u>25</u>

5.

7. 0, 7, 14, 11, 18, 25, 22, 29, <u>36, 33</u>

9. If a triangle has three equal angles, then it is equiangular.
    This triangle has three equal angles; therefore, it is equiangular.   VALID

11. If Jill sleeps too late, she will be late for class.
    Jill is not late for class; therefore, she did not sleep too late.   VALID

13. All high school students take math.
    Mary takes math; therefore, Mary is a high school student.   NOT VALID

15. All rectangles are parallelograms.
    This figure is not a parallelogram; therefore, it is not a rectangle.   VALID

17. ○ ○ ○ ○ ○ ○ ○

19. ○ △ □ ○ △ □ ○

21. −3, 0, 5, 12, 21, <u>32</u>

23. 6, 8, 12, 18, <u>26</u>

25. All fishermen using worms for bait caught fish.
    Joe caught a fish; therefore, Joe used worms for bait.   NOT VALID

27. All asparagus is grown in Washington.
    Marge loves asparagus; therefore, Marge lives in Washington.   NOT VALID

29. All corn grown in Kansas is at least six feet tall.
    Thelma's corn grows to a height of six feet; therefore,
    Thelma's corn was grown in Kansas.   NOT VALID

31. All students must take writing to graduate.
    Jenny did graduate; therefore, Jenny took writing.   VALID

33. 78, 63, 48, 33, <u>18</u>

35. 1, 3, 7, 15, 31, <u>63</u>

37. 1, 2, 3, 5, 8, 13, <u>21</u>

39. All triangles have three sides.
    Polygon ABC has three sides; therefore, polygon ABC is
    triangle.   DEDUCTIVE

41. A student observes that all prime numbers end in 1, 3, 5, 7, 9.
    She concludes that the number 1003 is a prime number.   INDUCTIVE

43. After measuring several swinging weights, Galileo concluded
    that the time of the swing of a pendulum is directly related to the
    square root of the length of the pendulum.   INDUCTIVE

45. On various occasions Kendall has burned his dinner while
    studying. He notices that chicken, carrots, beans, and
    other foods turn black if they are heated for a long time.
    He concludes that anything heated will eventually turn black. INDUCTIVE

47. answers vary

49. 6, 9, 18, 99, <u>6660</u>, <u>43,053,381</u>

51. $\dfrac{5.25 \text{ km}}{1} \cdot \dfrac{1000 \text{ m}}{\text{km}} = 5250 \text{ m}$

53. 7 g 88 mg + 2 g 95 mg = 9 g 183 mg =
    $9 \text{ g} + \dfrac{183 \text{ mg}}{1} \cdot \dfrac{\text{g}}{1000 \text{ mg}} = 9 \text{ g} + 0.183 \text{ g} = 9.183 \text{ g}$

55. $\dfrac{116.9 \text{ cm}}{5.6} = 20.875 \text{ cm}$

Section G.2
Geometry of Angles

1. 80° is an acute angle

3. 98° is an obtuse angle

5. 149° is an obtuse angle

7. 4° is an acute angle

9. 135° is an obtuse angle

11. The acute angles are: 18°, 45°, 82°

13. The reflex angles are: 210°, 315°, 282°

15. The obtuse angles are: 95°, 121°, 175°

17. Complement of 40° = 90° − 40° = 50°

19. Complement of 85° = 90° − 85° = 5°

21. Complement of 15° = 90° − 15° = 75°

23. Supplement of 80° = 180° − 80° = 100°

25. Supplement of 25° = 180° − 25° = 155°

27. Supplement of 55° = 180° − 55° = 125°

29. $\angle A = 180° − 134° = 46°$

31. $\angle C = 90° − 15° = 75°$

33. True

35. False − The supplement of an acute angle is an obtuse angle.

37. False − The sum of two acute angles may be an acute, a right, or an obtuse angle.

39. False − The sum of an obtuse angle and an acute angle may be an obtuse or
    a reflex angle.

41. True

43. When the two equations are subtracted from one another, $\angle B - \angle C = 0$, so then $\angle B = \angle C$

45. If $\angle B$ and $\angle C$ are complementary and $\angle C = 36°$, then $\angle B = 90° - 36° = 54°$. $\angle A = 180° - \angle B = 180° - 54° = 126°$.

47. An obtuse angle is greater than 90° and less than 180°.

49. Given: $\angle BAD = \angle ABD$ $\angle BAD = \dfrac{1}{2} \angle BAC$ $\angle ABD = \dfrac{1}{2} \angle ABC$

    By substitution: $\dfrac{1}{2} \angle BAC = \dfrac{1}{2} \angle ABC$

    Multiply both sides of this equation by 2 to show that $\angle BAC = \angle ABC$

51. $\dfrac{450\,\text{mi}}{2\,\text{hr}} \cdot \dfrac{5280\,\text{ft}}{\text{mi}} \cdot \dfrac{\text{hr}}{3600\,\text{sec}} = \dfrac{330\,\text{ft}}{\text{sec}}$

53. $\dfrac{17{,}824\,\text{in}}{1} \cdot \dfrac{\text{ft}}{12\,\text{in}} = 1485.3\,\text{ft}$

55. The perimeter is: $P = (28 + 42 + 16)\text{cm} \cdot \dfrac{\text{m}}{100\,\text{cm}} = 70.16\,\text{m}$

Section G.3
Geometry of Triangles

1. ΔABC is an isosceles triangle, since two sides are equal.

3. ΔABC is an equilateral triangle, since all three sides are equal.

5. ΔABC is an obtuse triangle, since $\angle C$ is greater than 90°.

7. ΔDEF is a right triangle, since $\angle E = 90°$.

9. ΔPQR is an obtuse triangle, since ∠R = is greater than 90°.
    ∠R = $180° - (45 + 30)° = 180° - 75° = 105°$

11. ΔRST is an obtuse triangle, since ∠S is greater than 90°.
    ∠T = $180° - (28 + 124)° = 180° - 152° = 28°$

13. ∠R = $180° - (45 + 30)° = 180° - 75° = 105°$

15. ∠T = $180° - (26 + 124)° = 180° - 150° = 30°$

17. ∠C = $180° - (20.3 + 94.7)° = 180° - 115° = 65°$

19. ∠D = $180° - (129.6 + 37.5)° = 180° - 167.1° = 12.9°$

21. True

23. False – The angles of equilateral triangles are all 60°, which is not an obtuse angle.

25. False – Some acute triangles could be isosceles triangles.

27. False – No isosceles triangles are scalene triangles.

29. True

31. True

33. answers vary

35. Given ∠2 and ∠3 are complementary and that ∠3 = ∠5.
    ΔABC is an obtuse triangle
    ΔCAE is a right triangle
    ΔADE is an acute triangle
    ΔABE is a right triangle
    ΔDBE is an obtuse triangle

37. Given ∠CAB = ∠CBA   and   ∠1 = ∠2
    Subtracting the second equation from the first equation gives ∠3 = ∠4.
    If two angles of a triangle are equal, then it is an isosceles triangle.
    This means that $\overline{AP} = \overline{BP}$

39. The triangle is a right triangle.

$$a^2 + b^2 \neq c^2$$
$$(16)^2 + (28)^2 \neq (42)^2$$
$$256 + 784 \neq 1764$$
$$1040 \neq 1784$$

41. The area of the square is: $A = s^2 = (23\,\text{m})^2 = 529\,\text{m}^2$

Section G.4
Congruent Triangles

1. a. To prove congruency by ASA, we need to show that $\angle I = \angle L$
    b. To prove congruency by SAS, we need to show that $\overline{GH} = \overline{JK}$
    c. To prove congruency by SSS, we need to show that $\overline{GH} = \overline{JK}$

3. $\triangle DAB \cong \triangle BCD$ by ASA

5. $\triangle AOC \cong \triangle BOD$ by SAS

7. $\triangle ADC \cong \triangle BEC$ by SAS

9. $\triangle ACD \cong \triangle BCE$ by ASA

11. $\triangle ABD \cong \triangle CBD$ by SAS

13. $\triangle ADC \cong \triangle BDC$ by SAS

15. True

17. Not necessarily true

19. $\overline{DF} = 12$ in.

21. $\triangle DEP \cong \triangle CEP$ by SAS or $\triangle APD \cong \triangle BPC$ by SAS

23. $\triangle ACE \cong \triangle BCD$ by SAS

25. $\triangle AED \cong \triangle BFC$ by SAS

27. $\triangle CEB \cong \triangle BFC$ by SAS

29. $\triangle ADN \cong \triangle CBM$ by ASA

31. answers vary

33. $\angle OBD = \angle OCD$ because $\triangle OBD \cong \triangle OCD$ by SAS

35. $\overline{AD} = \overline{AE}$ because $\triangle ACD \cong \triangle ACE$ by SAS

37. $\dfrac{120 \text{ words}}{\min} \cdot \dfrac{\min}{60 \text{ sec}} \cdot \dfrac{8 \text{ sec}}{1} = \dfrac{16 \text{ words}}{\text{sec}}$

39. The area of the circle is:
$$A = \pi r^2 = (3.14)(6.75 \text{ km})^2 = (3.14)(45.5625 \text{ km}^2) = 143.06625 \text{ km}^2$$

Section G.5
Parallel Lines

1. The pairs of alternate interior angles are: $2 - 8$ and $1 - 7$.

3. The interior angles are: 1, 2, 7, and 8.

5. $\angle 1$ and $\angle 4$ are vertical angles. Vertical angles are equal.
$\angle 1 = \angle 4 = 120°$

7. $\angle 1$ and $\angle 6$ are supplementary angles.
If $\angle 6 = 65°$, then $\angle 1 = 180° - 65° = 115°$

9. $\angle 6 = \angle 2 = \angle 5 = \angle 8$

11. $\angle 4$ is supplementary to $\angle 2$, $\angle 5$, $\angle 6$, $\angle 8$

13. $\angle 6 + \angle 3 + \angle 5 = 180°$ (straight angle)
$\angle 6 = \angle 1$ and $\angle 5 = \angle 2$ (alternate interior angles)
Therefore: $\angle 1 + \angle 3 + \angle 2 = 180°$ (substitution)

15. AB ∥ CD because line *l* is a transversal.
    ∠1 = ∠2 (given) and ∠2 = ∠3 (vertical angles).
    Therefore, ∠1 = ∠3 (corresponding angles)

17. ∠EFB = ∠KCD because:
    ∠FBC = ∠GCB (given)
    ∠FBC = ∠EFB (alternate interior angles) and ∠GCB = ∠KCD (vertical angles)
    Therefore, ∠EFB = ∠KCD (substitution)

19. answers vary

21. $\overline{AB} = \overline{CD}$ since ΔAOB ≅ ΔCOD by ASA

    Therefore, ∠1 = ∠2

23. $\overline{DG} \parallel \overline{EF}$ because ∠2 = ∠EFD (alternate interior angles)
    ∠1 = ∠2 (given) and ∠1 = ∠EFD (base angles of an isosceles triangle are equal)
    Therefore, ∠2 = ∠EFD

25. The volume of the box is:
$$V = (5\,\text{cm})(8\,\text{cm})(6\,\text{cm}) = 240\,\text{cm}^3 \cdot \left(\frac{\text{m}}{100\,\text{cm}}\right)^3 = 240\,\text{cm}^3 \cdot \frac{\text{m}^3}{1{,}000{,}000\,\text{cm}^3} = 0.00024\,\text{m}^3$$

27. The volume of the cone is:
$$V = \frac{1}{3}Bh = \frac{1}{3}\pi r^2 h = \frac{1}{3}(3.14)\left(8\,\text{in} \cdot \frac{\text{ft}}{12\,\text{in}}\right)^2 (1\,\text{ft}) =$$
$$\frac{1}{3}(3.14)\left(\frac{2}{3}\,\text{ft}\right)^2 (1\,\text{ft}) = \frac{1}{3}(3.14)\left(\frac{4}{9}\,\text{ft}^2\right)(1\,\text{ft}) \approx 0.47\,\text{ft}^3$$

Section G.6
Similar Triangles

1. $\angle E = \angle B = 180° - (\angle A + \angle C) = 180° - (46° + 76°) = 180° - 122° = 58°$

3. $\angle F = \angle C = 76°$

5. $\dfrac{\overline{FE}}{\overline{CB}} = \dfrac{\overline{DF}}{\overline{AC}} \Rightarrow \dfrac{\overline{FE}}{10} = \dfrac{15}{12} \Rightarrow 12\overline{FE} = 10(15) \Rightarrow 12\overline{FE} = 150 \Rightarrow \overline{FE} = 12.5$

7. $\dfrac{\overline{DC}}{\overline{AC}} = \dfrac{\overline{ED}}{\overline{BA}} \Rightarrow \dfrac{\overline{DC}}{\overline{DC}+6} = \dfrac{8}{20} \Rightarrow 20\overline{DC} = 8\overline{DC} + 48 \Rightarrow 12\overline{DC} = 48 \Rightarrow \overline{DC} = 4$

9. $\angle B = \angle CED = 40°$

11. $\dfrac{\overline{BD}}{\overline{AB}} = \dfrac{\overline{BC}}{\overline{AC}} \Rightarrow \dfrac{\overline{BD}}{13} = \dfrac{12}{5} \Rightarrow 5\overline{BD} = 12(13) \Rightarrow 5\overline{BD} = 156 \Rightarrow \overline{BD} = 31.2$

13. $\overline{DC} = \overline{AD} - \overline{AC} = 33.8 - 5 = 28.8$

15. $\dfrac{\overline{AC}}{\overline{BC}} = \dfrac{\overline{BC}}{\overline{CD}} \Rightarrow \dfrac{\overline{AC}}{9} = \dfrac{9}{12} \Rightarrow 12\overline{AB} = 9(9) \Rightarrow 12\overline{AB} = 81 \Rightarrow \overline{AB} = 6.75$

17. $\dfrac{\overline{AC}}{\overline{BC}} = \dfrac{\overline{AB}}{\overline{CD}} \Rightarrow \dfrac{\overline{AC}}{24} = \dfrac{21}{18} \Rightarrow 18\overline{AC} = 21(24) \Rightarrow 18\overline{AC} = 504 \Rightarrow \overline{AC} = 28$

19. $\dfrac{\overline{AC}}{\overline{BC}} = \dfrac{\overline{BC}}{\overline{BD}} \Rightarrow \dfrac{\overline{AC}}{120} = \dfrac{120}{90} \Rightarrow 90\overline{AC} = 120(120) \Rightarrow 90\overline{AC} = 14,400 \Rightarrow \overline{AC} = 160$

21. $\dfrac{\overline{AC}}{\overline{DF}} = \dfrac{\overline{AB}}{\overline{DE}} \Rightarrow \dfrac{8}{12} = \dfrac{4}{6} \Rightarrow 8(6) = 4(12) \Rightarrow 48 = 48$

$\dfrac{\overline{AC}}{\overline{DF}} = \dfrac{\overline{BC}}{\overline{EF}} \Rightarrow \dfrac{8}{12} = \dfrac{6}{9} \Rightarrow 12(6) = 8(9) \Rightarrow 72 = 72$

$\dfrac{\overline{AB}}{\overline{DE}} = \dfrac{\overline{BC}}{\overline{EF}} \Rightarrow \dfrac{4}{6} = \dfrac{6}{9} \Rightarrow 6(6) = 4(9) \Rightarrow 36 = 36$

23. $\dfrac{\overline{AD}}{\overline{BD}} = \dfrac{\overline{BD}}{\overline{CD}} \Rightarrow \dfrac{9}{12} = \dfrac{12}{16} \Rightarrow 9(16) = 12(12) \Rightarrow 144 = 144$

$\dfrac{\overline{BD}}{\overline{CD}} = \dfrac{\overline{AB}}{\overline{BC}} \Rightarrow \dfrac{12}{16} = \dfrac{15}{20} \Rightarrow 12(20) = 15(16) \Rightarrow 240 = 240$

$\dfrac{\overline{AD}}{\overline{BD}} = \dfrac{\overline{AB}}{\overline{BC}} \Rightarrow \dfrac{9}{12} = \dfrac{15}{20} \Rightarrow 12(15) = 20(9) \Rightarrow 180 = 180$

25. The approximate height of the rock formation is:

$\dfrac{5\,\text{in}}{75\,\text{ft}} = \dfrac{12\,\text{in}}{h} \Rightarrow 5h = 12(75\,\text{ft}) \Rightarrow 5h = 900\,\text{ft} \Rightarrow h = 180\,\text{ft}$

27. The lengths of the two unknown sides are:

$\dfrac{\overline{EF}}{\overline{BC}} = \dfrac{\overline{DE}}{\overline{AB}} \Rightarrow \dfrac{34\,\text{ft}}{40\,\text{ft}} = \dfrac{\overline{DE}}{52\,\text{ft}} \Rightarrow 40\overline{DE} = 34(52\,\text{ft}) \Rightarrow$

$40\overline{DE} = 1768\,\text{ft} \Rightarrow \overline{DE} = 44.2\,\text{ft}$

$\dfrac{\overline{EF}}{\overline{BC}} = \dfrac{\overline{DF}}{\overline{AC}} \Rightarrow \dfrac{34\,\text{ft}}{40\,\text{ft}} = \dfrac{\overline{DF}}{34\,\text{ft}} \Rightarrow 40\overline{DF} = 34(34\,\text{ft}) \Rightarrow$

$40\overline{DF} = 1156\,\text{ft} \Rightarrow \overline{DF} = 28.9\,\text{ft}$

The perimeter of the plot of ground is $(34 + 44.2 + 28.9)\text{ft} = 107.1\,\text{ft}$

29. $\dfrac{x}{15} = \dfrac{10}{12} \Rightarrow 12x = 15(10) \Rightarrow 12x = 150 \Rightarrow x = 12.5$

31. $\Delta CDE \sim \Delta CAB$ because:

$\angle DFE = \angle DFE$ (any angle is equal to itself)

$\dfrac{\overline{FH}}{\overline{FD}} = \dfrac{\overline{FK}}{\overline{FE}} \Rightarrow \dfrac{15}{20} = \dfrac{18}{24} \Rightarrow 15(24) = 20(18) \Rightarrow 360 = 360$

Therefore,

$\dfrac{\overline{FH}}{\overline{FD}} = \dfrac{\overline{HK}}{\overline{DE}} \Rightarrow \dfrac{15}{20} = \dfrac{\overline{HK}}{16} \Rightarrow 20\overline{HK} = 15(16) \Rightarrow 20\overline{HK} = 240 \Rightarrow \overline{HK} = 12$

33. $\Delta CDE \sim \Delta CDE$ because:

$\angle ACB = \angle DCE$ (vertical angles are equal)

$\angle CAB = \angle DEC$ (alternate interior angles are equal)

$\angle CBA = \angle CDE$ (alternate interior angles are equal)

35. ∆ABC ~ ∆DEA because:

    ∠BAC = ∠DAE (vertical angles are equal)

    ∠ACB = ∠ADE (alternate interior angles are equal)

    ∠ABC = ∠AED (alternate interior angles are equal)

37. answers vary

39. ∆ABD ~ ∆ABC because:

$$\left(\overline{AB}\right)^2 = \left(\overline{BC}\right)\left(\overline{BD}\right) = \frac{\overline{AB}}{\overline{BC}} = \frac{\overline{BD}}{\overline{AB}}$$ which means the corresponding sides are

    proportional.

41. Evaluate:

$$Q = -6x + 3y$$
$$Q = -6(3) + 3(-1)$$
$$Q = -18 - 3$$
$$Q = -21$$

43. $x = -5$ is not a solution:

$$3x - 45 - 2x \neq 50$$
$$3(-5) - 45 - 2(-5) \neq 50$$
$$-15 - 45 + 10 \neq 50$$
$$-60 + 10 \neq 50$$
$$-50 \neq 50$$

Plane Geometry
True-False Concept Review

1. True

2. True

3. True

4. True

5. False - 60° + 30° = 90°

6. False – An isosceles triangle could also be obtuse with angles of 20°, 20°, and 140°

7. False – The triangles are similar.

8. False – The sides and the angles must be in corresponding positions.

9. False – The alternate interior angles are on different sides of the transversal.

10. True

11. True

12. False – The triangles are similar.

13. False – All congruent triangles are similar triangles.

14. True

15. False – People under 21 years old might have a social security number.

16. False – The supplement has a measure of 45°.

17. True

18. False – A scalene triangle has no equal sides

19. False – A right triangle cannot be an acute triangle because it contains one right ∠

20. True

21. True

22. False – Vertical angles are equal.

23. False – Alternate interior angles are equal.

24. True

25. True

Plane Geometry
Review

1. 5, 11, 17, 23, <u>29</u>, <u>35</u>

3. 5, 3, 0, 7, 16, 14, 11, 18, 27, <u>25</u>, <u>22</u>

5. All chickens from the Value Deli are grown in Oregon.
   Pete bought chicken for dinner from the Value Deli; therefore,
   Pete brought an Oregon grown chicken.   **VALID**

7. The reflex angles are: 210° , 346°

9. The straight angles is: 180°

11. $\angle A = 90° - 55° = 35°$

13. $\angle S = 180° - 178° = 2°$

15. False – The supplement of a right angle is another right angle.

17. $\triangle ABC$ is a scalene triangle.

19. False – Every equilateral triangle is an isosceles triangle.

21. $\triangle ABC$ is an acute triangle because the third angle is less than 90°.

23. $\triangle DEF$ is an right triangle because the third angle is a right angle.

25. $\angle Q = 180° - (110° + 25°) = 180° - 135° = 45°$

27. The measures of $\angle A$ and $\angle D$ are needed to prove congruency by SAS.

29. $\triangle ABE$ and $\triangle CBE$ are congruent by ASA

31. $\angle A = \angle R$

33. $\overline{BC} = \overline{ST}$

35. $\overline{RT} = \overline{AC}$

37. $\angle 1$ and $\angle 4$ form a pair of vertical angles.

39. ∠7 and ∠3 form a pair of corresponding angles.

41. ∠1 = 135°

43. ∠2 = 45°

45. ∠3 = 45°

47. $\dfrac{\overline{BC}}{\overline{AB}} = \dfrac{\overline{CD}}{\overline{AC}} \Rightarrow \dfrac{35}{37} = \dfrac{\overline{CD}}{12} \Rightarrow 37\overline{CD} = 35(12) \Rightarrow 37\overline{CD} = 420 \Rightarrow \overline{CD} = 11\dfrac{13}{37}$

49. The height of the power pole is:
$$\dfrac{\overline{AB}}{\overline{AD}} = \dfrac{\overline{AC}}{\overline{AE}} \Rightarrow \dfrac{16}{16+64} = \dfrac{12}{\overline{AE}} \Rightarrow \dfrac{16}{80} = \dfrac{12}{\overline{AE}} \Rightarrow$$
$$16\overline{AE} = 12(80) \Rightarrow 16\overline{AE} = 960 \quad \overline{AE} = 60\,\text{ft}$$

51. To show that ΔBEC ~ ΔAED:
    ∠BEC = ∠AED (vertical angles are equal)
    ∠CBE = ∠EDA and ∠BCE = ∠EAD (alternate interior angles are equal)

53. To show that ΔAEC ~ ΔCDB:
    ∠AEC = ∠CDB (right angles are equal)
    ΔABC is an isosceles triangle since $\overline{AB} = \overline{AC}$
    ∠ABC = ∠ACB (the base angles of an isosceles triangle are equal)

55. ΔABD ~ ΔBDC since:
$$\dfrac{\overline{AB}}{\overline{BC}} = \dfrac{\overline{AD}}{\overline{CD}} \Rightarrow \dfrac{7.5}{10} = \dfrac{4.5}{6} \Rightarrow 10(4.5) = 6(7.5) \Rightarrow 45 = 45$$
$$\dfrac{\overline{AB}}{\overline{BC}} = \dfrac{\overline{BD}}{\overline{CD}} \Rightarrow \dfrac{7.5}{10} = \dfrac{6}{8} \Rightarrow 10(6) = 8(7.5) \Rightarrow 60 = 60$$
$$\dfrac{\overline{AD}}{\overline{BD}} = \dfrac{\overline{BD}}{\overline{CD}} \Rightarrow \dfrac{4.5}{6} = \dfrac{6}{8} \Rightarrow 8(4.5) = 6(6) \Rightarrow 36 = 36$$

Plane Geometry
Test

1.

2. True

3. ∠2 and ∠7 are exterior angles
    ∠2 and ∠4 are vertical angles
    ∠3 and ∠6 are interior angles

4. ∠3 and ∠8 are supplementary angles. ∠3 = 180° − ∠8 = 180° − 75° = 105°.

5. 180° − (67° + 75.8°) = 180° − 142.8° = 37.2°

6. ΔDEF is a scalene triangle.

7. The two triangles are similar with the AAA proof.

8. ∠D = 90° − 48° = 42°

9. False

10. The obtuse angles are: 125°, 100°

11. The two triangles are congruent with the SAS proof.

12. All students must have 2.00 or better GPA to graduate.
    June had a 3.76 GPA; therefore, June graduated.   NOT VALID

13. The interior angles on the same side of the transversal are supplementary.

14. $\overline{AF} = \overline{FC}$

15. The length of $\overline{DE}$ is:
    $$\frac{\overline{AC}}{\overline{DF}} = \frac{\overline{AB}}{\overline{DE}} \Rightarrow \frac{75}{15} = \frac{60}{\overline{DE}} \Rightarrow 75 \cdot \overline{DE} = 60(15) \Rightarrow 75 \cdot \overline{DE} = 900 \Rightarrow \overline{DE} = 12$$